Trigonometry

Trigonometry

Eric Bradford

STATES
ACADEMIC PRESS
www.statesacademicpress.com

Published by States Academic Press
109 South 5th Street,
Brooklyn, NY 11249, USA
www.statesacademicpress.com

Trigonometry
Eric Bradford

International Standard Book Number: 978-1-63989-532-8 (Hardback)

Cataloging-in-Publication Data

Trigonometry / Eric Bradford.
p. cm.
Includes bibliographical references and index.
ISBN 978-1-63989-532-8
1. Trigonometry. 2. Geometry. 3. Mathematics. I. Bradford, Eric.
QA531 .T75 2022
516.24--dc23

Table of Contents

Preface VII

Chapter 1 What is Trigonometry? 1
- Trigonometric Ratios 6
- Trigonometric Substitution 8
- Trigonometric Polynomials 15
- Trigonometry Angles 20
- Trigonometric Integrals 26
- Applications 29

Chapter 2 Trigonometric Functions: An Integrated Study 33
- Sine 35
- Cosine 50
- Tangent 63
- Cosecant 70
- Secant 80
- Cotangent 90

Chapter 3 Understanding Inverse Trigonometric Functions 102
- Basic Concepts of Inverse Trigonometric Functions 102
- Properties of Inverse Trigonometric Functions 105
- Arcsine 109
- Arccosine 113
- Arctangent 115
- Arccosecant 119
- Arcsecant 121
- Arccotangent 125

- Proofs of Derivatives of Inverse Trigonometric Functions 129
- Relationships Among the Inverse Trigonometric Functions 133

Chapter 4 Trigonometric Identities and Formulas 135
- Hermite's Cotangent Identity 167
- Morrie's Law 167
- Pythagorean Trigonometric Identity 169
- Proofs of Trigonometric Identities 174
- Half Angle Formula 187
- Double Angle Formulas 194
- Triple Angle Formulas 199

Chapter 5 Trigonometry Laws 203
- Law of Sines 203
- Law of Cosines 208
- Law of Tangents 227

Permissions

Index

Preface

It is with great pleasure that I present this book. It has been carefully written after numerous discussions with my peers and other practitioners of the field. I would like to take this opportunity to thank my family and friends who have been extremely supporting at every step in my life.

The field of mathematics that focuses on the relationship between angles and side lengths of triangles is known as trigonometry. The real functions that relates an angle of the right-angled triangle with the ratios of the two side lengths are known as trigonometric functions. This field is known for its triangle and trigonometric identities, which are used to solve complex equations and find new relationships. Some of the triangle identities are law of sines, cosines and tangents whereas Pythagorean identities and Euler's formula are some of the trigonometric identities. This field finds its applications in the fields of optics, acoustics, astronomy, celestial mechanics, periodic functions, navigation, surveying, phonetics, geodesy, architecture, economics, computer graphics, crystallography, electrical engineering, biology and medical imaging. This book is a valuable compilation of topics, ranging from the basic to the most complex theories and principles related to this discipline. Different approaches, evaluations, methodologies and advanced studies related to this field have been included in this book. Through this book, we attempt to further enlighten the readers about the new concepts in trigonometry.

The chapters below are organized to facilitate a comprehensive understanding of the subject:

Chapter - What is Trigonometry?

The branch of mathematics that deals with the study of relationships between side lengths and angles of triangles is called trigonometry. They can be represented using the unit circle and are used in the areas of celestial mechanics, optics and acoustics, geodesy and navigation. The topics elaborated in this chapter will help in gaining a better perspective of trigonometry.

Chapter - Trigonometric Functions: An Integrated Study

Trigonometric functions are the angle functions between an angle of a right-angled triangle and ratios of two side lengths. It includes sine, cosine, tangent, cosecant, secant and cotangent. This chapter has been carefully written to provide an easy understanding of the trigonometric functions.

Chapter - Understanding Inverse Trigonometric Functions

Inverse trigonometric functions are the inverse functions of sine, cosine, tangent, cotangent, secant, and cosecant functions. They can be used for obtaining an angle from any of the angle's trigonometric ratios. All the concepts related to inverse trigonometric functions have been carefully analyzed in this chapter.

Chapter – Trigonometric Identities and Formulas

Hermite's cotangent identity, Morrie's law, Pythagorean identities, double angle formulas, half-angle formulas, triple-angle formulas of sine, cosine and tangent, etc. are studied within trigonometric identities. This chapter closely examines these trigonometric identities and formulas to provide an extensive understanding of the subject.

Chapter – Trigonometry Laws

There are various laws that are studied under the domain of trigonometry. These are law of sines, cosines which includes spherical and hyperbolic law of cosines, and tangents. This chapter has been carefully written to provide an easy understanding of these trigonometry laws.

Eric Bradford

1

What is Trigonometry?

The branch of mathematics that deals with the study of relationships between side lengths and angles of triangles is called trigonometry. They can be represented using the unit circle and are used in the areas of celestial mechanics, optics and acoustics, geodesy and navigation. The topics elaborated in this chapter will help in gaining a better perspective of trigonometry.

Trigonometry is a branch of mathematics that deals with triangles, particularly those plane triangles in which one angle has 90 degrees (right triangles). Trigonometry deals with relationships between the sides and the angles of triangles and with the trigonometric functions, which describe those relationships.

Trigonometry has applications in both pure mathematics and in applied mathematics, where it is essential in many branches of science and technology. It is usually taught in secondary schools either as a separate course or as part of a precalculus course. Trigonometry is informally called "trig" or "trigo."

A branch of trigonometry, called spherical trigonometry, studies triangles on spheres, and is important in astronomy and navigation.

The Canadarm2 robotic manipulator on the International Space Station is operated by controlling the angles of its joints. Calculating the final position of the astronaut at the end of the arm requires repeated use of the trigonometric functions of those angles.

If one angle of a triangle is 90 degrees and one of the other angles is known, the third is thereby fixed, because the three angles of any triangle add up to 180 degrees. The two acute angles therefore add up to 90 degrees: They are complementary angles. The shape of a right triangle is completely determined, up to similarity, by the angles. This means that once one of the other angles is known, the ratios of the various sides are always the same regardless of the overall size of the triangle. These ratios are given by the following trigonometric functions of the known angle A, where a, b, and c refer to the lengths of the sides in the accompanying figure:

- The sine function (sin), defined as the ratio of the side opposite the angle to the hypotenuse.

$$\sin A = \frac{\text{opposite}}{\text{hypotenuse}} = \frac{a}{c}.$$

- The cosine function (cos), defined as the ratio of the adjacent leg to the hypotenuse.

$$\cos A = \frac{\text{adjacent}}{\text{hypotenuse}} = \frac{b}{c}.$$

- The tangent function (tan), defined as the ratio of the opposite leg to the adjacent leg.

$$\tan A = \frac{\text{opposite}}{\text{adjacent}} = \frac{a}{b} = \frac{\sin A}{\cos A}.$$

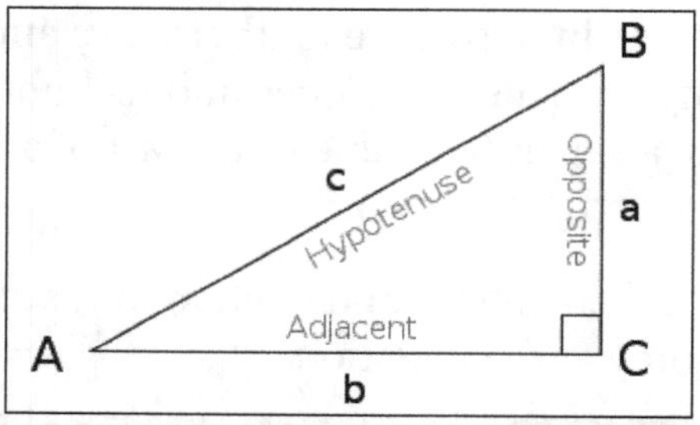

In this right triangle: sin A = a/c; cos A = b/c; tan A = a/b.

The hypotenuse is the side opposite to the 90 degree angle in a right triangle; it is the longest side of the triangle, and one of the two sides adjacent to angle A. The adjacent leg is the other side that is adjacent to angle A. The opposite side is the side that is opposite to angle A. The terms perpendicular and base are sometimes used for the opposite and adjacent sides respectively. Many people find it easy to remember what sides of the right triangle are equal to sine, cosine, or tangent, by memorizing the word SOH-CAH-TOA.

The reciprocals of these functions are named the cosecant (csc or cosec), secant (sec) and cotangent (cot), respectively. The inverse functions are called the arcsine, arccosine, and arctangent, respectively. There are arithmetic relations between these functions, which are known as trigonometric identities.

With these functions one can answer virtually all questions about arbitrary triangles by using the law of sines and the law of cosines. These laws can be used to compute the remaining angles and sides of any triangle as soon as two sides and an angle or two angles and a side or three sides are known. These laws are useful in all branches of geometry, since every polygon may be described as a finite combination of triangles.

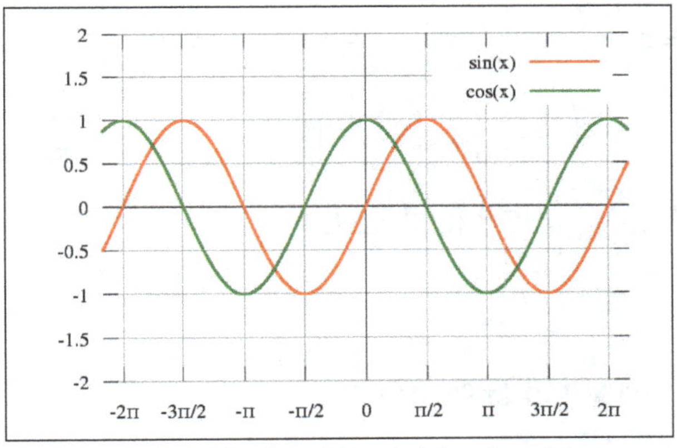

Graphs of the functions sin(x) and cos(x), where the angle x is measured in radians.

The above definitions apply to angles between 0 and 90 degrees (0 and $\pi/2$ radians) only. Using the unit circle, one can extend them to all positive and negative arguments (see trigonometric function). The trigonometric functions are periodic, with a period of 360 degrees or 2π radians. That means their values repeat at those intervals.

The trigonometric functions can be defined in other ways besides the geometrical definitions above, using tools from calculus and infinite series. With these definitions the trigonometric functions can be defined for complex numbers. The complex function cis is particularly useful:

$$\operatorname{cis} x = \cos x + i \sin x = e^{ix}.$$

Common Formulae

Certain equations involving trigonometric functions are true for all angles and are known as trigonometric identities. Many express important geometric relationships. For example, the Pythagorean identities are an expression of the Pythagorean Theorem. Here are some of the more commonly used identities, as well as the most important formulae connecting angles and sides of an arbitrary triangle.

Trigonometric Identities

Pythagorean Identities:

$$\sin^2\alpha + \cos^2\alpha = 1$$

$$\tan^2\alpha + 1 = \sec^2\alpha$$

$$1 + \cot^2\alpha = \csc^2\alpha$$

Sum and Product Identities

Sum to Product:

$$\sin\alpha \pm \sin\beta = 2\sin\left(\frac{\alpha\pm\beta}{2}\right)\cos\left(\frac{\alpha\mp\beta}{2}\right)$$

$$\cos\alpha + \cos\beta = 2\cos\left(\frac{\alpha+\beta}{2}\right)\cos\left(\frac{\alpha-\beta}{2}\right)$$

$$\cos\alpha - \cos\beta = -2\sin\left(\frac{\alpha+\beta}{2}\right)\sin\left(\frac{\alpha-\beta}{2}\right)$$

Product to Sum:

$$\cos\alpha\cos\beta = \frac{1}{2}[\cos(\alpha-\beta)+\cos(\alpha+\beta)]$$

$$\sin\alpha\sin\beta = \frac{1}{2}[\cos(\alpha-\beta)-\cos(\alpha+\beta)]$$

$$\cos\alpha\sin\beta = \frac{1}{2}[\sin(\alpha+\beta)-\sin(\alpha-\beta)]$$

$$\sin\alpha\cos\beta = \frac{1}{2}[\sin(\alpha+\beta)+\sin(\alpha-\beta)]$$

Sine, Cosine and Tangent of a Sum

$$\sin(\alpha\pm\beta) = \sin\alpha\cos\beta \pm \cos\alpha\sin\beta$$

$$\cos(\alpha\pm\beta) = \cos\alpha\cos\beta \mp \sin\alpha\sin\beta$$

$$\tan(\alpha\pm\beta) = \frac{\tan\alpha \pm \tan\beta}{1\mp\tan\alpha\tan\beta}$$

Half Angle Identities

Note that $\pm$ is correct, it means it may be either one, depending on the value of A/2.

$$\sin\frac{A}{2} = \pm\sqrt{\frac{1-\cos A}{2}}$$

$$\cos\frac{A}{2} = \pm\sqrt{\frac{1+\cos A}{2}}$$

$$\tan\frac{A}{2} = \pm\sqrt{\frac{1-\cos A}{1+\cos A}} = \frac{\sin A}{1+\cos A} = \frac{1-\cos A}{\sin A}$$

Stereographic (or Parametric) Identities

$$\sin\alpha = \frac{2T}{1+T^2}$$

$$\cos\alpha = \frac{1-T^2}{1+T^2}$$

where,

$$T = \tan\frac{\alpha}{2}.$$

Triangle Identities

In the following identities, A, B, and C are the angles of a triangle and a, b, and c are the lengths of sides of the triangle opposite the respective angles.

Law of Sines

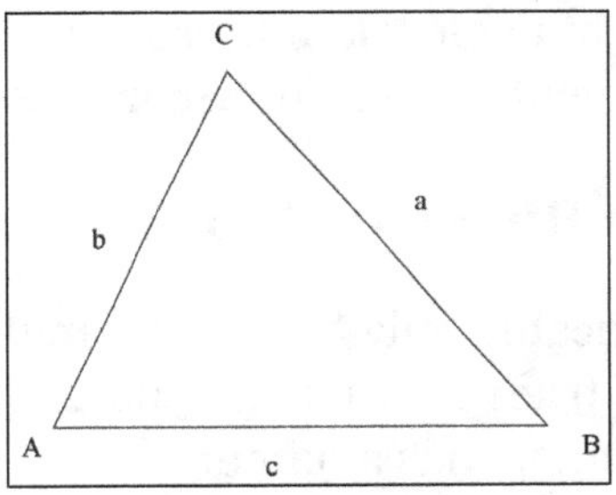

Laws of Sines.

The law of sines (also know as the "sine rule") for an arbitrary triangle states:

$$\frac{a}{\sin A} = \frac{b}{\sin B} = \frac{c}{\sin C} = 2R,$$

where R is the radius of the circumcircle of the triangle.

Law of Cosines

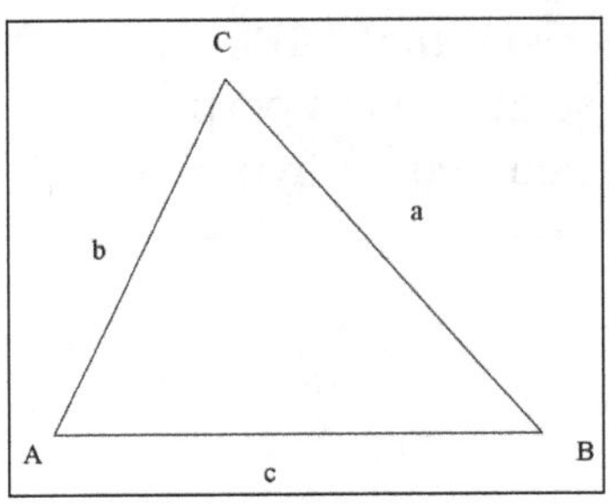

Laws of Cosines.

The law of cosines (also known as the cosine formula, or the "cos rule") is an extension of the Pythagorean theorem to arbitrary triangles:

$$c^2 = a^2 + b^2 - 2ab\cos C,$$

or equivalently:

$$\cos C = \frac{a^2 + b^2 - c^2}{2ab}.$$

Law of Tangents

$$\frac{a+b}{a-b} = \frac{\tan\left[\frac{1}{2}(A+B)\right]}{\tan\left[\frac{1}{2}(A-B)\right]}$$

TRIGONOMETRIC RATIOS

Six trigonometric ratios for right angle triangle are Sin, Cos, Tan, Cosec, Sec, Cot which stands for Sine, Cosecant, Tangent, Cosecant, Secant respectively.

Trigonometric Ratios in Right Angle Triangle

Trigonometric Ratios are applicable only for a right angle triangle. It is a triangle with specialty, that one angle of the triangle will be of 900 and rest two will be less than 900. Each side of the right angle triangle has a name:

- Hypotenuse: Like its name, it is the largest side of the triangle. In other words, it is the largest side opposite to the right angle of the triangle.
- Base: Side on which right angle triangle stands is known as its base. Any of the two sides other than the hypotenuse can be chosen as the base for performing the calculation.
- Perpendicular: The side perpendicular to the base the right-angled triangle will be perpendicular to the right-angled triangle.

Consider a right angle triangle right angle ABC. Let at C be of 90°. Side AB will be hypotenuse then, chose AC as base and BC as perpendicular. Follow the given figure: for angle BAC, value of sin θ = Perpendicular/hypotenuse = BC/AB.

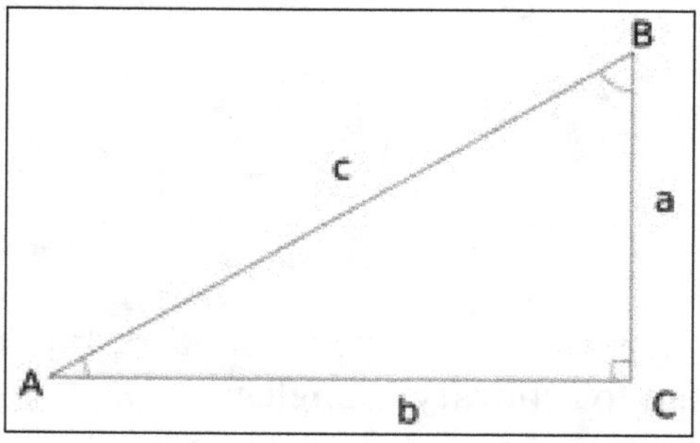

Right Angle Triangle ABC.

Concepts of Trigonometric Ratios

Fixing the base and perpendicular can be difficult sometimes. For example in the triangle above:

- For angle BAC, $\sin\theta_1$ = Perpendicular/ Hypotenuse = BC/AB.
- But for angle ABC, $\sin\theta_2$ = Perpendicular/ Hypotenuse = AC/AB.
- Confusing, isn't? To remove this confusion, we will name different sides of the right-angled triangle as adjacent, opposite and hypotenuse.
- Adjacent: It will be the side adjacent to the angle being considered.
- Opposite: It will be the side opposite to angle being considered.
- Hypotenuse: It will be the side opposite to the right angle of the triangle (or the largest side).

Now, formulas for ratios are as follows:

- Sin θ = Perpendicular/Hypotenuse = Opposite/Hypotenuse.
- Cosine or cos θ = Base/Hypotenuse = Adjacent/Hypotenuse.
- Tangent or tan θ = Perpendicular/Base = Opposite/Adjacent.

The reciprocal of sin, cos, and cot can also have names and it's obvious they are also trigonometric ratios. Those are as follows:

cosec θ = Hypotenuse/Perpendicular = Hypotenuse/Opposite

sec θ = Hypotenuse/Base = Hypotenuse/Adjacent

cot θ = Base/Perpendicular = Adjacent/Opposite

Trigonometric Ratios Table

angle / ratio	0°	30°	45°	60°	90°
sin θ	0	$\frac{1}{2}$	$\frac{1}{\sqrt{2}}$	$\frac{\sqrt{3}}{2}$	1
cos θ	1	$\frac{\sqrt{3}}{2}$	$\frac{1}{\sqrt{2}}$	$\frac{1}{2}$	0
tan θ	0	$\frac{1}{\sqrt{3}}$	1	$\sqrt{3}$	not defined
cosec θ	not defined	2	$\sqrt{2}$	$\frac{2}{\sqrt{3}}$	1

sec θ	1	$\frac{2}{\sqrt{3}}$	$\sqrt{2}$	2	not defined
cot θ	not defined	$\sqrt{3}$	1	$\frac{1}{\sqrt{3}}$	0

Example: For a right angle triangle ABC right angled at C with Hypotenuse=AB = 5cm, Perpendicular = BC = 4cm and Base = AC = 3cm. Calculate for angle BAC = θ value of sin θ, cos θ and tan θ.

Solution: For right angle triangle ABC,

Sin θ = Perpendicular/ Hypotenuse = 4/5

Cos θ = Base/ Hypotenuse = 3/5

Tan θ = Perpendicular/Base = 4/3

TRIGONOMETRIC SUBSTITUTION

In finding the area of a circle or an ellipse, an integral of the form $\int\sqrt{a^2-x^2}\,dx$ arises, where a > 0. If it were $\int\sqrt{a^2-x^2}\,dx$, the substitution u = $a^2 - x^2$ would be effective but, as it stands, $\int\sqrt{a^2-x^2}\,dx$ is more difficult. If we change the variable from to by the substitution $x = a\sin\theta$, then the identity $1-\sin^2\theta=\cos^2\theta$ allows us to get rid of the root sign because:

$$\sqrt{a^2-x^2}=\sqrt{a^2-a^2\sin^2\theta}=\sqrt{a^2\left(1-\sin^2\theta\right)}=\sqrt{a^2\cos^2\theta}=a\left|\cos\theta\right|$$

Notice the difference between the substitution u = $a^2 - x^2$ (in which the new variable is a function of the old one) and the substitution $x = a\sin\theta$ (the old variable is a function of the new one).

In general we can make a substitution of the form x = g(t) by using the Substitution Rule in reverse. To make our calculations simpler, we assume that g has an inverse function; that is, g is one-to-one. In this case, if we replace u by x and x by t in the Substitution Rule, we obtain:

$$\int f(x)dx=\int f(g(t))g'(t)dt$$

This kind of substitution is called inverse substitution.

We can make the inverse substitution $x = a\sin\theta$ provided that it defines a one-to-one function. This can be accomplished by restricting θ to lie in the interval $[-\pi/2,\pi/2]$.

In the following table we list trigonometric substitutions that are effective for the given radical expressions because of the specified trigonometric identities. In each case the restriction on θ is imposed to ensure that the function that defines the substitution is one-to-one.

Table: Trigonometric Substitutions.

Expression	Substitution	Identity
$\sqrt{a^2 - x^2}$	$x = a\sin\theta,\ -\frac{\pi}{2} \le \theta \le \frac{\pi}{2}$	$1 - \sin^2\theta = \cos^2\theta$
$\sqrt{a^2 + x^2}$	$x = a\tan\theta,\ -\frac{\pi}{2} < \theta < \frac{\pi}{2}$	$1 + \tan^2\theta = \sec^2\theta$
$\sqrt{x^2 - a^2}$	$x = a\sec\theta,\ 0 \le \theta < \frac{\pi}{2} \text{ or } \pi \le \theta \le \frac{3\pi}{2}$	$\sec^2\theta - 1 = \tan^2\theta$

Example: Evaluate $\int \frac{\sqrt{9 - x^2}}{x^2} dx$

Solution: Let, $x = 3\sin\theta$,

where, $-\pi/2 \le \theta \le \pi/2$.

Then,

$$dx = 3\cos\theta\, d\theta \text{ and } \sqrt{9 - x^2} = \sqrt{9 - 9\sin^2\theta} = \sqrt{9\cos^2\theta} = 3\cos\theta$$

($\cos\theta \ge 0$ because $-\pi/2 \le \theta \le \pi/2$). Thus, the Inverse Substitution Rule gives:

$$\begin{aligned}\int \frac{\sqrt{9 - x^2}}{x^2} dx &= \int \frac{3\cos\theta}{3\sin^2\theta} 3\cos\theta\, d\theta \\ &= \int \frac{\cos^2\theta}{\sin^2\theta} d\theta = \int \cot^2\theta\, d\theta \\ &= \int \left(\csc^2\theta - 1\right) d\theta \\ &= -\cot\theta - \theta + C\end{aligned}$$

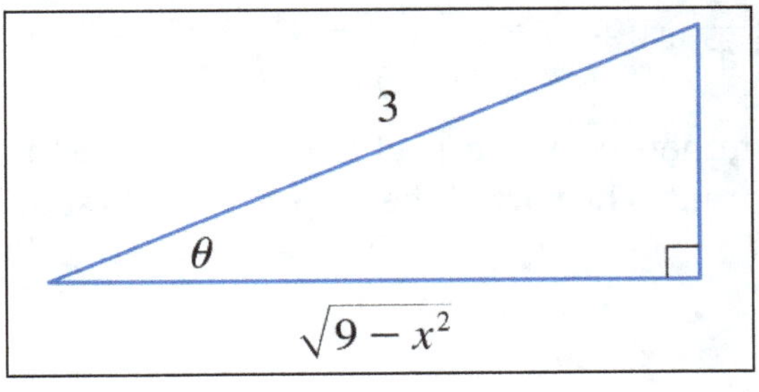

$\text{Sin}\,\theta = \frac{x}{3}.$

Since this is an indefinite integral, we must return to the original variable x. This can be done either by using trigonometric identities to express cot θ in sin θ = x/3 terms of or by drawing a diagram, where is interpreted as an angle of a right triangle. Since sin θ = x/3, we label the opposite side and the hypotenuse as having lengths x and 3. Then the Pythagorean Theorem gives the length of the adjacent side as $\sqrt{9-x^2}$, so we can simply read the value of cot θ from the figure:

$$\cot\theta = \frac{\sqrt{9-x^2}}{x}$$

(Although θ > o in the diagram, this expression for cot θ is valid even when θ < o.) Since sin θ = x/3, we have θ = $\sin^{-1}$ (x/3) and so:

$$\int \frac{\sqrt{9-x^2}}{x^2}dx = -\frac{\sqrt{9-x^2}}{x} - \sin^{-1}\left(\frac{x}{3}\right) + C$$

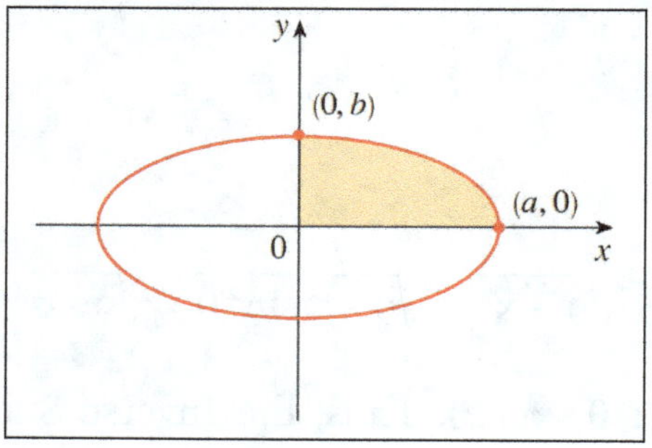

$$\frac{x^2}{a^2} + \frac{y^2}{b^2} = 1.$$

Example: Find the area enclosed by the ellipse:

$$\frac{x^2}{a^2} + \frac{y^2}{b^2} = 1$$

Solution: Solving the equation of the ellipse for y, we get:

$$\frac{y^2}{b^2} = 1 - \frac{x^2}{a^2} = \frac{a^2 - x^2}{a^2} \quad \text{or} \quad y = \pm\frac{b}{a}\sqrt{a^2 - x^2}$$

Because the ellipse is symmetric with respect to both axes, the total area A is four times the area in the first quadrant. The part of the ellipse in the first quadrant is given by the function:

$$y = \frac{b}{a}\sqrt{a^2 - x^2} \qquad 0 \le x \le a$$

And so,

$$\frac{1}{4}A = \int_0^a \frac{b}{a}\sqrt{a^2 - x^2}\,dx$$

To evaluate this integral we substitute x = a sin θ. Then dx = a cos θ dθ. To change the limits of integration we note that when x = 0, sin θ = 0, so θ = 0; when x = a, sin θ = 1, so θ = π/2. Also,

$$\sqrt{a^2 - x^2} = \sqrt{a^2 - a^2\sin^2\theta} = \sqrt{a^2\cos^2\theta} = a\left|\cos\theta\right| = a\cos\theta$$

Since, $0 \le \theta \le \pi/2$. Therefore,

$$A = 4\frac{b}{a}\int_0^a \sqrt{a^2 - x^2}\,dx = 4\frac{b}{a}\int_0^{\pi/2} a\cos\theta \cdot a\cos\theta\,d\theta$$

$$= 4ab\int_0^{\pi/2}\cos^2\theta\,d\theta = 4ab\int_0^{\pi/2}\frac{1}{2}(1 + \cos 2\theta)\,d\theta$$

$$= 2ab\left[\theta\ \frac{1}{2}\sin 2\theta\right]_0^{\pi/2} = 2ab\left(\frac{\pi}{2} + 0 - 0\right)$$

$$= \pi ab$$

We have shown that the area of an ellipse with semiaxes a and b is πab. In particular, taking a = a = r, we have proved the famous formula that the area of a circle with radius r is πr^2.

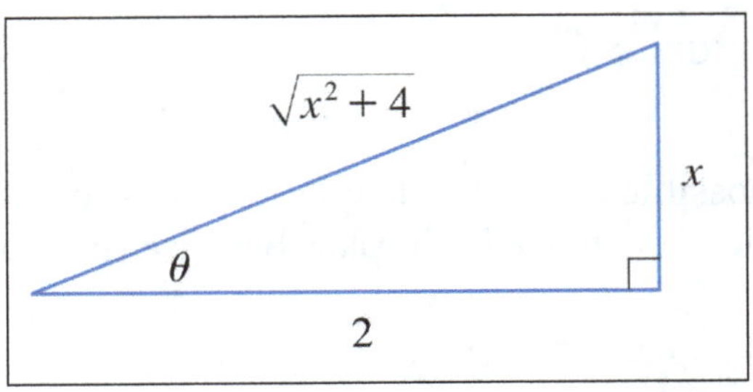

$$\tan\theta = \frac{x}{2}.$$

Example: Find $\int \frac{1}{x^2\sqrt{x^2 + 4}}dx$

Solution: Let, $x = 2\tan\theta, -\pi/2 < \theta < \pi/2$. Then, $dx = 2\sec^2\theta\,d\theta$ and

$$\sqrt{x^2+4} = \sqrt{4(\tan^2\theta + 1)} = \sqrt{4\sec^2\theta} = 2\left|\sec\theta\right| = 2\sec\theta$$

Thus, we have:

$$\int \frac{dx}{x^2\sqrt{x^2+4}} = \int \frac{2\sec^2\theta\, d\theta}{4\tan^2\theta \cdot 2\sec\theta} = \frac{1}{4}\int \frac{\sec\theta}{\tan^2\theta} d\theta$$

To evaluate this trigonometric integral we put everything in terms of sin θ and cos θ:

$$\frac{\sec\theta}{\tan^2\theta} = \frac{1}{\cos\theta}\cdot\frac{\cos^2\theta}{\sin^2\theta} = \frac{\cos\theta}{\sin^2\theta}$$

Therefore, making the substitution u = sin θ, we have:

$$\int \frac{dx}{x^2\sqrt{x^2+4}} = \frac{1}{4}\int \frac{\cos\theta}{\sin^2\theta} d\theta = \frac{1}{4}\int \frac{du}{u^2}$$

$$= \frac{1}{4}\left(-\frac{1}{u}\right) + C = -\frac{1}{4\sin\theta} + C = -\frac{\csc\theta}{4} + C$$

We use figure to determine that:

$$\csc\theta = \sqrt{x^2+4}/x$$

and so,

$$\int \frac{dx}{x^2\sqrt{x^2+4}} = -\frac{\sqrt{x^2+4}}{4x} + C$$

Example: Find $\int \frac{x}{\sqrt{x^2+4}} dx$

Solution: It would be possible to use the trigonometric substitution x = 2 tan θ, here. But the direct substitution u = x^2 + 4 is simpler, because then du = 2x dx and

$$\int \frac{x}{\sqrt{x^2+4}} dx = \frac{1}{2}\int \frac{du}{\sqrt{u}} = \sqrt{u} + C = \sqrt{x^2+4} + C$$

Example: $\int \frac{dx}{\sqrt{x^2-a^2}}$, where a>0

Solution: We let, x = a sec θ, where 0 < θ <π/2 or π < θ < 3π/2. Then, dx = a sec θ tan θ d θ and

$$\sqrt{x^2-a^2} = \sqrt{a^2(\sec^2\theta-1)} = \sqrt{a^2\tan^2\theta} = a|\tan\theta| = a\tan\theta$$

Therefore,

$$\int \frac{dx}{\sqrt{x^2 - a^2}} = \int \frac{a \sec\theta \tan\theta}{a \tan\theta} d\theta$$

$$= \int \sec\theta \, d\theta = \ln|\sec\theta + \tan\theta| + C$$

The triangle in figure gives $\tan\theta = \sqrt{x^2 - a^2} / a$, so we have:

$$\int \frac{dx}{\sqrt{x^2 - a^2}} = \ln\left|\frac{x}{a} + \frac{\sqrt{x^2 - a^2}}{a}\right| + C$$

$$= \ln\left|x + \sqrt{x^2 - a^2}\right| - \ln a + C$$

Writing $C_1 = C - \ln a$, we have:

$$\int \frac{dx}{\sqrt{x^2 - a^2}} = \ln\left|x + \sqrt{x^2 - a^2}\right| + C_1$$

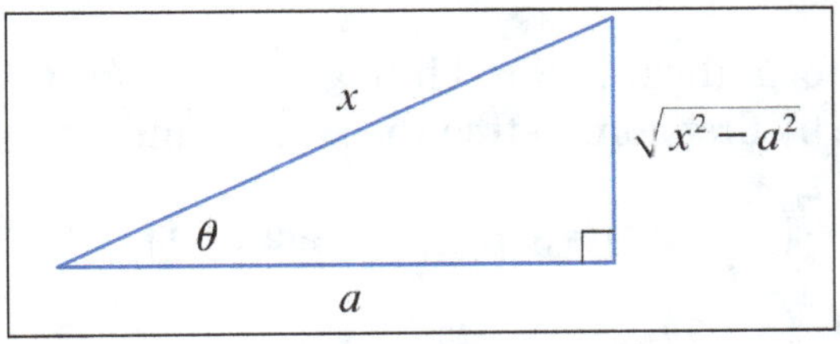

$$\sec\theta = \frac{x}{a}.$$

Example: Find $\int_0^{3\sqrt{3}/2} \frac{x^3}{(4x^2 + 9)^{3/2}} dx$

Solution: First we note that $(4x^2 + 9)^{3/2} = \left(\sqrt{4x^2 + 9}\right)^3$ so trigonometric substitution is appropriate. Although $\sqrt{4x \quad 9}$ is not quite one of the expressions in the table of trigonometric substitutions, it becomes one of them if we make the preliminary substitution u = 2x. When we combine this with the tangent substitution, we have x = 3/2 tan θ, which gives dx = 3/2 sec² θ dθ and

$$\sqrt{4x^2 + 9} = \sqrt{9\tan^2\theta + 9} = 3\sec\theta$$

When x = 0, tan θ = 0, so θ = 0; when $x = 3\sqrt{3}/2$, $\tan\theta = \sqrt{3}$, so $\theta = \pi/3$.

$$\int_0^{3\sqrt{3}/2} \frac{x^3}{(4x^2 + 9)^{3/2}} dx = \int_0^{\pi/3} \frac{\frac{27}{8}\tan^3\theta}{27\sec^3\theta} \frac{3}{2}\sec^3\theta \, d\theta$$

$$= \frac{3}{16}\int_0^{\pi/3} \frac{\tan^3\theta}{\sec\theta}\,d\theta = \frac{3}{16}\int_0^{\pi/3} \frac{\sin^3\theta}{\cos^2\theta}\,d\theta$$

$$= \frac{3}{16}\int_0^{\pi/3} \frac{1-\cos^2}{\cos^2\theta}\sin\theta\,d\theta$$

Now we substitute u = cos θ so that du = –sin θ dθ. When θ = 0, u = 1; when θ = π/3, u = ½.

Therefore,

$$\int_0^{3\sqrt{3}/2} \frac{x^3}{\left(4x^2+9\right)^{3/2}}\,dx = -\frac{3}{16}\int_1^{1/2} \frac{1-u^2}{u^2}\,du = \frac{3}{16}\int_1^{1/2}\left(1-u^{-2}\right)du$$

$$= \frac{3}{16}\left[u+\frac{1}{u}\right]_1^{1/2} = \frac{3}{16}\left[\left(\frac{1}{2}+2\right)-(1+1)\right] = \frac{3}{32}$$

Example: Evaluate $\int \frac{x}{\sqrt{3-2x-x^2}}\,dx$

Solution: We can transform the integrand into a function for which trigonometric substitution is appropriate by first completing the square under the root sign:

$$3-2x-x^2 = 3-\left(x^2+2x\right) = 3+1-\left(x^2+2x+1\right)$$
$$= 4-(x+1)^2$$

This suggests that we make the substitution u = x + 1. Then, du = dx and x = u – 1, so,

$$\int \frac{x}{\sqrt{3-2x-x^2}}\,dx = \int \frac{u-1}{\sqrt{4-u^2}}\,du$$

We now substitute u = 2 sin θ, giving du = 2 cos θ dθ and $\sqrt{4-u^2} = 2\cos\theta$, so,

$$\int \frac{x}{\sqrt{3-2x-x^2}}\,dx = \int \frac{2\sin\theta-1}{2\cos\theta}2\cos\theta\,d\theta$$

$$= \int (2\sin\theta-1)\,d\theta$$

$$= -2\cos\theta-\theta+C$$

$$= -\sqrt{4-u^2}-\sin^{-1}\left(\frac{u}{2}\right)+C$$

$$= -\sqrt{3-2x-x^2}-\sin^{-1}\left(\frac{x+1}{2}\right)+C$$

TRIGONOMETRIC POLYNOMIALS

By a trigonometric polynomial we mean a function of the form:

$$f(t) = \sum_{n=o}^{N} (a_n \cos nt + b_n \sin nt)$$

where a_n and b_n are constants which allowed to be complex. One can think of such a function as a combination of several harmonics (described by sinusoidal functions sin nt and cos nt) for a musical instrument.

An alternative description of this function is:

$$f(t) = \sum_{n=-N}^{N} c_n e^{int}$$

Indeed, Using $\cos nt = (e^{int} + e^{-int})/2$ and $\sin nt = (e^{int} - e^{-int})/2i$, we can convert equation $f(t) = \sum_{n=o}^{N} (a_n \cos nt + b_n \sin nt)$ into the form of equation $f(t) = \sum_{n=-N}^{N} c_n e^{int}$. Using $e^{\pm int} = \cos nt \pm i \sin nt$ we can turn equation $f(t) = \sum_{n=-N}^{N} c_n e^{int}$ back to equation $f(t) = \sum_{n=o}^{N} (a_n \cos nt + b_n \sin nt)$. We prefer to use equation $f(t) = \sum_{n=-N}^{N} c_n e^{int}$ because,

$$\left\{e^{int}\right\} - \infty < n < \infty = \left\{ \ldots, e^{-2it}, e^{-it}, 1, e^{it}, e^{2it}, \ldots \right\}$$

is an orthonormal system with respect to the inner product:

$$\langle f, g \rangle = \frac{1}{2\pi} \int_0^{2\pi} f(t) \overline{g(t)} dt.$$

As we know, the general theory of orthonormal systems tells us that the coefficient c_n in equation $f(t) = \sum_{n=-N}^{N} c_n e^{int}$ is given by the following recipe:

$$c_n = \left\langle f, e^{int} \right\rangle \equiv \frac{1}{2\pi} \int_0^{2\pi} f(t) e^{-int} dt.$$

Here, we use the method of harmonic analysis to derive Cauchy's formula for polynomials and Poisson formula for trigonometric polynomials, which will be extended in the future for more general functions. Both formulas are extremely important one for complex analysis and the other for harmonic analysis. The basic tactic in our approach is to exploit the orthonormality of $\left\{e^{int}\right\} - \infty < n < \infty$. Before we start, we show two (related) examples in order help the reader to appreciate this tactic.

Example: We wish to find the integral:

$$I_m = \frac{1}{2\pi}\int_o^{2\pi} \cos^m t\,dt$$

for any positive integer m. When m is odd, clearly I_m = o. So we may assume m to be even, say m = 2n. Now:

$$\cos^m t = \left(\frac{e^{it}+e^{-it}}{2}\right)^m = \frac{1}{2^m}\sum_{k=o}^{m}\binom{m}{k} e^{i(m-k)t_e-ikt} = \frac{1}{2^m}\sum_{k=o}^{m}\binom{m}{k} e^{i(m-2k)t}.$$

From the above discussion, we have:

$$\frac{1}{2\pi}\int_o^{2\pi} e^{i(m-2k)t}dt \frac{1}{2\pi}\int_o^{2\pi} e^{i(2n-2k)t}dt = \begin{cases} 2\pi & \text{if } k=n; \\ o & \text{otherwise.} \end{cases}$$

So, writing m = 2n, we have:

$$\frac{1}{2\pi}\int_o^{2\pi} \cos^{2n} t\,dt = \frac{1}{2\pi}\int_o^{2\pi}\sum_{k=o}^{2n}\binom{2n}{k} e^{i(2n-2k)t}dt = \frac{1}{2^{2n}}\binom{2n}{n}.$$

Example: We have established that $\{e^{int}\}_{n\epsilon Z}$ is an ONS (orthonormal system).

The identity,

$$\cos^n t = \sum_{k=o}^{n}\frac{1}{2^n}\binom{n}{k} e^{i(n-2k)t}$$

can be viewed as an orthogonal decomposition with respect to this ONS. (Here we replace m by n. This identity holds without assuming m even). So we may apply Pythagoras identity to obtain,

$$\sum_{k=o}^{n}\left\{\frac{1}{2^n}\binom{n}{k}\right\}^2 = \left\|\cos^n t\right\|^2 \equiv \frac{1}{2\pi}\int_o^{2\pi}\cos^{2n} t\,dt$$

which is $\binom{2n}{n}/2^{2n}$, according to the last example. So we obtain,

$$\sum_{k=o}^{n}\binom{n}{k}^2 = \binom{2n}{n}$$

which is neat and interesting identity.

Take a polynomial function,

$$p(z) = a_o + a_1 z + a_2 z^2 + \ldots + a_N z^N \equiv \sum\nolimits_{n=o}^{N} a_n z^n$$

in a complex variable z. For convenience, we ignore N and rewrite it as,

$$p(z) = \sum_{n\ge o} a_n z^n$$

with the understanding that $a_n = o$ for n > N. Let us restrict it to the unit circle by considering the function $f(t) = p(e^{it})$. Then,

$$f(t) \equiv p(e^{it}) = \sum_{n\ge o} a_n e^{int}$$

This is a trigonometric polynomial and its coefficients c_n are given by,

$$a_n = \langle f, e_n \rangle \equiv \frac{1}{2\pi}\int_o^{2\pi} f(t) e^{-int} dt.$$

Fix a point z in the complex plane lying inside the unit circle, i.e. |z| < 1. Then,

$$p(z) = \sum_{n\ge o} a_n z^n = \sum_{n\ge o} \langle f, e_n \rangle z^n$$
$$= \sum_{n\ge o} \langle f, \bar{z}^n e_n \rangle \equiv \sum_{n\ge o} \frac{1}{2\pi}\int_o^{2\pi} f(t) z^n e^{-int} dt$$
$$= \frac{1}{2\pi}\int_o^{2\pi} \sum_{n\ge o} f(t) z^n e^{-int} dt = \frac{1}{2\pi}\int_o^{2\pi} f(t) \sum_{n\ge o} z^n e^{-int} dt$$

$\sum_{n\ge o} z^n e^{-int}$ can be viewed as a geometric series and hence we can evaluate its sum of functions as follows:

$$\sum_{n\ge o} z^n e^{-int} = \sum_{n\ge o} \left(ze^{-int}\right)^n = \frac{1}{1-ze^{-it}}.$$

(Notice that $\left|ze^{-it}\right| = |z| < 1$ and, when $|w| < 1$, $\text{sum}_{n\ge o}\ w^n = 1/(1-w)$.) Therefore, when $|z| < 1$, the above equation gives,

$$p(z) = \frac{1}{2\pi}\int_o^{2\pi} \frac{f(t)}{1-ze^{-it}} dt \equiv \frac{1}{2\pi}\int_o^{2\pi} \frac{p(e^{it})}{1-ze^{-it}} dt.$$

We would like to put this identity in a neat way. For this purpose, we introduce a new variable w by putting $w = e^{it}$. Then |w| = 1, that is, w varies along the unit circle. Notice that, when t runs from 0 to 2π, $w = e^{it}$ runs from 1 along the unit circle counterclockwisely once and returns to 1 at the end. Formally, we can differentiate w to get,

$$dw = d\left(e^{it}\right) = \frac{d\left(e^{it}\right)}{dt} dt = ie^{it}\, dt = iw\, dt$$

and hence dt = dw/iw. Thus the integrand on the right hand side of equation $p(z) = \frac{1}{2\pi}\int_o^{2\pi} \frac{f(t)}{1-ze^{-it}} dt \equiv \frac{1}{2\pi}\int_o^{2\pi} \frac{p(e^{it})}{1-ze^{-it}} dt$ becomes,

$$\frac{p(e^{it})}{1-ze^{-it}} dt = \frac{p(w)}{1-zw^{-1}} \frac{dw}{iw} = \frac{1}{i}\frac{p(w)}{w-z} dw.$$

Hence we can rewrite equation $p(z) = \frac{1}{2\pi}\int_0^{2\pi}\frac{f(t)}{1-ze^{-it}}dt \equiv \frac{1}{2\pi}\int_0^{2\pi}\frac{p(e^{it})}{1-ze^{-it}}dt$ as,

$$p(z) = \frac{1}{2\pi i}\int_{|w|=1}\frac{p(w)}{w-z}dw.$$

The above identity is a special case of the celebrated Cauchy's formula in complex analysis. More precisely, it may be called the Cauchy formula for polynomial functions over the unit circle.

By putting z = 0 in equation $p(z) = \frac{1}{2\pi}\int_0^{2\pi}\frac{f(t)}{1-ze^{-it}}dt \equiv \frac{1}{2\pi}\int_0^{2\pi}\frac{p(e^{it})}{1-ze^{-it}}dt$ we obtain:

$$p(0) = \frac{1}{2\pi}\int_0^{2\pi}p(e^{it})dt.$$

The right hand side is the average value of a polynomial p over the unit circle, which is equal to the value of p at the center of the circle. We call this the mean value property.

Equation $p(z) = \frac{1}{2\pi i}\int_{|w|=1}\frac{p(w)}{w-z}dw$ is actually valid for any circle. To see this, let C be a circle centered at z_0 with radius r:

$$C = \{z \in \mathbf{Z} : |z - z_0| = r\}.$$

The usual parametric equation for this circle is,

$$z = z_0 + re^{it}, \quad 0 \le t \le 2\pi.$$

Given a polynomial function p(z), consider the function $q(\zeta) = p(z_0 + r\zeta)$, which is another polynomial function of the same degree in variable ζ. We can apply equation $p(z) = \frac{1}{2\pi}\int_0^{2\pi}\frac{f(t)}{1-ze^{-it}}dt \equiv \frac{1}{2\pi}\int_0^{2\pi}\frac{p(e^{it})}{1-ze^{-it}}dt$ to this polynomial to get,

$$p(z_0 + r\zeta) = \frac{1}{2\pi}\int_0^{2\pi}\frac{p(z_0 + re^{it})}{1-\zeta e^{-it}}dt.$$

For $|\zeta| < 1$. Now, take any point z inside the circle C, that is, $|z - z_0| < r$. We can write $z = z_0 + r\zeta$ for some ζ with $|\zeta| < 1$. Indeed, just let $\zeta = (z - z_0)/r$ from which it follows that $|\zeta| = |z - z_0|/r < 1$ in view of $|z - z_0| < r$. Thus the above equation gives,

$$p(z) = \frac{1}{2\pi}\int_0^{2\pi}\frac{p(z_0 + re^{it})}{1-(z-z_0)r^{-1}e^{-it}}dt = \frac{1}{2\pi}\int_0^{2\pi}\frac{p(z_0 + re^{it})re^{it}\,dt}{z_0 - re^{it} - z}.$$

Now the parametric equation for C, namely $w = z_o + re^{it}$ (here we cannot write $z = z_o + re^{it}$ because we have already used the letter z for a point inside the circle C), gives $dw = ire^{it}\ dt$. Hence the above equation becomes,

$$p(z) = \frac{1}{2\pi i}\int_{|w-z_o|=r} \frac{p(w)}{w-z}dw.$$

The real part of a polynomial,

$$p(z) = a_o + a_1 z + a_2 z^2 + \ldots + a_N z^N \equiv \sum_{n=o}^{N} a_n z^n$$

is given by Re $p(z) = (\overline{p(z)} + p(z))/2$. Putting $z = re^{it}$, and $f_r(t) = 2\mathbf{Re}\, p\left(re^{it}\right) \equiv \overline{p(e^{it})} + p\left(re^{it}\right)$, we have,

$$f_r(t) = \overline{p(re^{it})} + p\left(re^{it}\right) = \sum_{n=o}^{N} \bar{a}_n r^n e^{-int} + \sum_{n=o}^{N} a_n r^r e^{int}.$$

Let us rewrite this identity as,

$$f_r(t) = \sum_{n=-N}^{N} c_n r^{|n|} e^{int}$$

where $c_n = a_n$ for $n > o$, $c_n = \bar{a}_n$ for $n < o$, and $c_o = a_o + \bar{a}_o$. Notice that $|n| = n$ if $n \geq o$ and $|n| = -n$ if $n < o$. When r = 1, we simply write f for f_r. Thus,

$$f(t) = \sum_{n=-N}^{N} c_n e^{int}$$

The Fourier coefficient c_n is given by,

$$c_n = \frac{1}{2\pi}\int_o^{2\pi} f(s)e^{-ins}ds$$

Again, let us ignore the limits of the summation in equation $f_r(t) = \sum_{n=-N}^{N} c_n r^{|n|} e^{int}$ and write,

$$f_r(t) = \sum_{-\infty<n<\infty} c_n r^{|n|} e^{int} \equiv \cdots + c{-2}r^2 e^{-2it} + c_{-1} re^{-it} + c_o + c_1 re^{it} + c_2 r^2 e^{2it} + \cdots$$

by tacitly assuming that $c_n = o$ for $|n| > N$. Substituting the Fourier coefficients c_n given in equation $c_n = \frac{1}{2\pi}\int_o^{2\pi} f(s)e^{-ins}ds$ into equation,

$$f_r(t) = \sum_{-\infty<n<\infty} c_n r^{|n|} e^{int} \equiv \cdots + c{-2}r^2 e^{-2it} + c_{-1} re^{-it} + c_o + c_1 re^{it} + c_2 r^2 e^{2it} + \cdots$$

we get,

$$f_r(t)=\sum_{-\infty<n<\infty}\left(\frac{1}{2\pi}\int_o^{2\pi} f(s)e^{-ins}\,ds\right)r^{|n|}e^{int}$$

$$=\frac{1}{2\pi}\int_o^{2\pi}\left(\sum_{-\infty<n<\infty} r^{|n|}e^{in(t-s)}\right)f(s)ds.$$

Here we have switched the integral and summation, This is allowed due to the uniform convergence of the series,

$$\sum_{-\infty<n<\infty} r^{|n|}e^{in(t-s)}.$$

Let $w=re^{i(t-s)}$. Then this series can be written as,

$$\cdots+\bar{w}^2+\bar{w}+1+w+w^2+\cdots$$

or

$W+\bar{W}$, where $W=\frac{1}{2}+\sum_{n=1}^{\infty} w^n$. Now:

$$W=\left(\sum_{n=o}^{\infty} w^n\right)-\frac{1}{2}=\frac{1}{1-w}-\frac{1}{2}=\frac{1}{2}\frac{1+w}{1-w}.$$

So,

$$W+\bar{W}=\frac{1}{2}\frac{1+w}{1-w}+\frac{1}{2}\frac{1+\bar{w}}{1-\bar{w}}=\frac{1}{2}\frac{(1+w)(1-\bar{w})+(1+\bar{w})(1-w)}{(1-w)(1-\bar{w})}$$

$$=\frac{1-w\bar{w}}{1-(w+\bar{w})+w\bar{w}}=\frac{1-r^2}{1-2r\cos(t-s)+r^2}$$

in view of $w\bar{w}=|w|^2=r^2$ and $w+\bar{w}=2\operatorname{Re}w=2r\cos(t-s)$. Substituting the last expression, we get,

$$f_r(t)=\frac{1}{2\pi}\int_o^{2\pi}\frac{1-r^2}{1-2r\cos(t-s)+r^2}f(s)ds$$

which is called the Poisson integral of f. For any continuous function of period 2π, not necessarily a trigonometric polynomial, f_r converges to f as r increases to 1.

TRIGONOMETRY ANGLES

The angles $m\pi/n$ (with m, n integers) for which the trigonometric functions may be expressed in terms of finite root extraction of real numbers are limited to values of m which are precisely those which produce constructible polygons. Analytic expressions for trigonometric functions with arguments of this form can be obtained using the Wolfram Language function, for values of $n \geq 7$ (for $n \leq 6$, the trigonometric functions auto-evaluate in the Wolfram Language).

Compass and straightedge constructions dating back to Euclid were capable of inscribing regular polygons of 3, 4, 5, 6, 8, 10, 12, 16, 20, 24, 32, 40, 48, 64, ..., sides. However, Gauss showed in 1796 (when he was 19 years old) that a sufficient condition for a regular polygon on n sides to be constructible was that n be of the form:

$$n = 2^k p1p2 \cdots ps,$$

where k is a nonnegative integer and the pi are distinct Fermat primes. Here, a Fermat prime is a prime Fermat number, i.e., a prime number of the form:

$$F_n = 2^{2^n} + 1,$$

where $n \geq 0$ is an integer, and the only known primes of this form are 3, 5, 17, 257, and 65537. The first proof of the fact that this condition was also necessary is credited to Wantzel.

A necessary and sufficient condition that a regular n-gon be constructible is that $\phi(n)$ be a power of 2, where $\phi(n)$ is the totient function,

Constructible values of n for n < 300 were given by Gauss, and the first few are 1, 2, 3, 4, 5, 6, 8, 10, 12, 15, 16, 17, 20, 24, 30, 32, 34, 40, 48, 51, 60, 64, 68, 80, 85, 96, 102, 120, 128, 136, 160, 170, 192, ...

The algebraic degrees of $\cos(2\pi/n)$ for constructible polygons are 1, 1, 1, 1, 2, 1, 2, 2, 2, 4, 4, 8, 4, 4, 4, 8, ..., and of $\cos(\pi/n)$ are:

1	1	1
1 1	1 1	3
1 2 1	1 0 1	5
1 3 3 1	1 1 1 1	15
1 4 6 4 1	1 0 0 0 1	17
1 5 10 10 5 1	1 1 0 0 1 1	51
1 6 15 20 15 6 1	1 0 1 0 1 0 1	85
1 7 21 35 35 21 7 1	1 1 1 1 1 1 1 1	255
1 8 28 56 70 56 28 8 1	1 0 0 0 0 0 0 0 1	257

Gardner and independently Watkins noticed that the number of sides for constructible polygons with odd Numbers of sides are given by the first 32 rows of the Sierpińs-ki sieve interpreted as binary numbers, giving 1, 3, 5, 15, 17, 51, 85, 255, ... In other words, every row is a product of distinct Fermat primes, with terms given by binary counting.

A partial table of the analytic values of sine, cosine, and tangent for arguments π/n with small integer n is given in table.

Derivations of these formulas appear in the following entries:

$x(°)$	$x(\text{rad})$	$\sin x$	$\cos x$	$\tan x$
0.0	0	0	1	0
15.0	$\frac{1}{12}\pi$	$\frac{1}{4}(\sqrt{6}-\sqrt{2})$	$\frac{1}{4}(\sqrt{6}+\sqrt{2})$	$2-\sqrt{3}$
18.0	$\frac{1}{10}\pi$	$\frac{1}{4}(\sqrt{5}-1)$	$\frac{1}{4}\sqrt{10+2\sqrt{5}}$	$\frac{1}{5}\sqrt{25-10\sqrt{5}}$
22.5	$\frac{1}{8}\pi$	$\frac{1}{2}\sqrt{2-\sqrt{2}}$	$\frac{1}{2}\sqrt{2+\sqrt{2}}$	$\sqrt{2}-1$
30.0	$\frac{1}{6}\pi$	$\frac{1}{2}$	$\frac{1}{2}\sqrt{3}$	$\frac{1}{3}\sqrt{3}$
36.0	$\frac{1}{5}\pi$	$\frac{1}{4}\sqrt{10-2\sqrt{5}}$	$\frac{1}{4}(1+\sqrt{5})$	$\sqrt{5-2\sqrt{5}}$
45.0	$\frac{1}{4}\pi$	$\frac{1}{2}\sqrt{2}$	$\frac{1}{2}\sqrt{2}$	1
60.0	$\frac{1}{3}\pi$	$\frac{1}{2}\sqrt{3}$	$\frac{1}{2}$	$\sqrt{3}$
90.0	$\frac{1}{2}\pi$	1	0	∞
180.0	π	0	-1	0

There is a nice mnemonic for remembering sines of common angles,

$$\sin(0°)=\frac{1}{2}\sqrt{0}$$

$$\sin(30°)=\frac{1}{2}\sqrt{1}$$

$$\sin(45°)=\frac{1}{2}\sqrt{2}$$

$$\sin(60°)=\frac{1}{2}\sqrt{3}$$

$$\sin(90°)=\frac{1}{2}\sqrt{4}.$$

In general, any trigonometric function can be expressed in radicals for arguments of the form $r\pi$, where $r=m/n$ is a rational number, by writing the trigonometric functions in exponential form and the exponentials as roots of -1. For example,

$$\cos\left(\frac{\pi}{23}\right)=-\frac{1}{2}(-1)^{22/23}\left[1+(-1)^{2/23}\right].$$

This confirms that for r rational, trigonometric functions of $r\pi$ are always algebraic numbers. For example, the cases n=7 and n=9 involve the cubic equation (in $\sin^2(\pi/7)$ and $\sin^2(\pi/9)$, respectively). The polynomial of which a given expression is a root can be obtained in the Wolfram Language using the syntax Root Reduce (To Radicals expr), which produces a Root object.

Letting $(P(x))_n$ denoted the nth root of the polynomial P(x) in the ordering of the Wolfram Language's Root object, the first few analytic values of $\sin(\pi/n)$ are summarized in the following table:

n	$\sin(\pi/n)$
1	0
2	1
3	$\frac{1}{2}\sqrt{3}$
4	$\frac{1}{2}\sqrt{2}$
5	$(16x^4-20x^2+5)_3$
6	$\frac{1}{2}$
7	$(64x^6-112x^4+56x^2-7)_4$
8	$(8x^4-8x^2+1)_3$
9	$(64x^6-9x^4+36x^2-3)_4$
10	$\frac{1}{4}(\sqrt{5}-1)$

The algebraic order of $\sin(\pi/n)$ is given analytically by,

$$\operatorname{ord}\left(\sin\left(\frac{\pi}{n}\right)\right)=\begin{cases}1 & \text{if } n=2\\ \phi(n) & \text{if } n\equiv 0,1,3 \pmod 4\\ \frac{1}{2}\phi(n) & \text{otherwise}\end{cases}$$

where $\phi(n)$ is the totient function. For n=1, 2, ..., this gives the sequence 1, 1, 2, 2, 4, 1, 6, 4, 6, 2, 10, 4, ...

The minimal polynomial for $\sin(2\pi/p)$ with p an odd prime is given by,

$$S_p(x)=\sum_{k=0}^{(p-1)/2}(-1)^k\binom{p}{2k+1}(1-x^2)^{(p-1)/2-k}x^{2k}$$

If $n \neq 4$ and $n = 2^r m$, the algebraic order of $\sin(2\pi/n)$ is given by,

$$\operatorname{ord}\left(\sin\left(\frac{2\pi}{n}\right)\right) = \begin{cases} \phi(n) & \text{if } r = 0,1 \\ \frac{1}{4}\phi(n) & \text{if } r = 2 \\ \frac{1}{2}\phi(n) & \text{if } r \geq 3 \end{cases}$$

This gives the sequence 1, 1, 2, 1, 4, 2, 6, 2, 6, 4, 10, 1, ...

The first few analytic values of $\cos(\pi/n)$ are summarized in the following table:

n	$\cos(\pi/n)$
1	-1
2	0
3	$\frac{1}{2}$
4	$\frac{1}{2}\sqrt{2}$
5	$\frac{1}{4}(1+\sqrt{5})$
6	$\frac{1}{2}\sqrt{3}$
7	$(8x^3-4x^2-4x+1)_3$
8	$(8x^4-8x^2+1)_4$
9	$(8x^3-6x-1)_3$
10	$(16x^4-20x^2+5)_4$

The algebraic order of $\cos(\pi/n)$ is given analytically by,

$$\operatorname{ord}\left(\cos\left(\frac{\pi}{n}\right)\right) = \begin{cases} 1 & \text{if } n = 1 \\ \phi(n) & \text{if } n \equiv 0 \pmod 2 \\ \frac{1}{2}\phi(n) & \text{if } n \equiv 1 \pmod 2, \end{cases}$$

where ϕ is the totient function. For n=1, 2, ..., this gives the sequence 1, 1, 1, 2, 2, 2, 3, 4, 3, 4, 5, 4, 6, ...

The algebraic order of $\cos(2\pi/n)$ is given analytically by,

$$\operatorname{ord}\left(\cos\left(\frac{2\pi}{n}\right)\right)=\begin{cases}1 & \text{if } n=1 \text{ or } 2\\ \frac{1}{2}\phi(n) & \text{if } n>2\end{cases}$$

giving the sequence 1, 1, 1, 1, 2, 1, 3, 2, 3, 2, 5, ...

For $\cos(2\pi/p)$ with p an odd prime, an explicit formula can be given for the minimal polynomial, namely,

$$\Psi(x)=\sum_{i=0}^{s}(-1)^{i}\sigma_i(2x)^{s-i},$$

where,

$$s\equiv\frac{1}{2}(p-1)$$

$$\sigma_{2k}\equiv(-1)^{k}\binom{s-k}{k}$$

$$\sigma_{2k-1}\equiv(-1)^{k}\binom{s-k}{k-1}$$

Watkins and Zeitlin showed that,

$$T_{s+1}(x)-T_{s}(x)=2^{s}\prod_{d/n}\Psi(x)$$

for $n=2s+1$ odd, where $T_n(x)$ is a Chebyshev polynomial of the first kind, and

$$T_{s+1}(x)-T_{s-1}(x)=2^{s}\prod_{d/n}\Psi_d(x)$$

for n=2 s even.

Beslin and de Angelis give the simpler form,

$$C_p(x)=S_p\left(\sqrt{\frac{1-x}{2}}\right),$$

where $S_p(x)$ is as defined above.

A special type of expansion in terms of radicals with real arguments can be obtained if n is a power of two times a product of distinct Fermat primes. For other values of n, the situation becomes more complicated. It is now no longer possible to express trigonometric functions in a form that they are expressed as real radicals, but a certain minimal representation still exists. The simplest nontrivial example is for n=7. The exact meaning of "minimal" is rather technical and is related to the Galois subgroups of certain cyclotomic polynomials. As it turns out, for n prime, the expansions are especially interesting and difficult, and higher order Galois group calculations are both difficult and time-consuming. For example, n=23 is a very difficult case and takes a long time to calculate. Some larger primes are easier again but the complexity grows with the size of the prime on average.

While individual trigonometric functions may require complicated representations at certain angles, there are general formulas for the products of these functions. For example,

$$\prod_{k=1}^{n}\cos\left(\frac{k\pi}{n}\right)=\begin{cases}0 & \text{if } n\equiv 0 \pmod 2\\ \dfrac{(-1)^{(n+1)/2}}{2^{2n-1}} & \text{if } n\equiv 1 \pmod 2\end{cases}$$

$$\prod_{k=1}^{n}\sin\left(\frac{k\pi}{n}\right)=0$$

$$\prod_{k=1}^{\lfloor n/2\rfloor}\cos\left(\frac{k\pi}{n}\right)=\begin{cases}0 & \text{if } n\equiv 0 \pmod 2\\ \dfrac{1}{2^{(n-1)/2}} & \text{if } n\equiv 1 \pmod 2\end{cases}$$

$$\prod_{k=1}^{\lfloor n/2\rfloor}\sin\left(\frac{k\pi}{n}\right)=\sqrt{\frac{n}{2^{n-1}}}.$$

The first few values of the latter for n=1, 2, ... are therefore 1, 1, 3/4, 1/2, 5/16, 3/16, ... Another example is the general case of Morrie's law,

$$2^k\prod_{j=0}^{k-1}\cos\left(2^j a\right)=\frac{\sin\left(2^k a\right)}{\sin a}.$$

TRIGONOMETRIC INTEGRALS

The trigonometric integrals are a family of integrals involving trigonometric functions:

1. Integrals of the form $\int\cos ax\cos bx dx, \int\sin ax\cos bx dx, \int\sin ax\sin bx dx$.

To evaluate integrals of products of sine and cosine with different arguments, we apply the identities:

$$\cos ax\cos bx=\frac{1}{2}[\cos(ax+bx)+\cos(ax-bx)]$$

$$\sin ax \cos bx = \frac{1}{2}[\sin(ax+bx)+\sin(ax-bx)]$$

$$\sin ax \sin bx = -\frac{1}{2}[\cos(ax+bx)-\cos(ax-bx)]$$

2. Integrals of the form $\int \sin^m x \cos^n x dx$.

We assume here that the powers m and n are non-negative integers.

To find an integral of this form, use the following substitutions:

- If m (the power of sine) is odd, we use the u–substitution $u = \cos x, du = -\sin x dx$ and the identity $\sin^2 x + \cos^2 x = 1$ to express the remaining even power of sine in u–terms.
- If n (the power of cosine) is odd, we use the u–substitution $u = \sin x, du = \cos x dx$ and the identity $\sin^2 x + \cos^2 x = 1$ to express the remaining even power of cosine in u–terms.
- If both powers m and n are even, we reduce the powers using the the double angle formulas $\sin^2 x = \frac{1-\cos 2x}{2}, \cos^2 x = \frac{1+\cos 2x}{2}$.

The integrals of type $\int \sin^n x dx$ and $\int \cos^n x dx$ can be evaluated by reduction formulas:

$$\int \sin^n x dx = -\frac{\sin^{n-1} x \cos x}{n} + \frac{n-1}{n}\int \sin^{n-2} x dx,$$

$$\int \cos^n x dx = -\frac{\cos^{n-1} x \sin x}{n} + \frac{n-1}{n}\int \cos^{n-2} x dx.$$

3. Integrals of the form $\int \tan^n x dx$.

The power of the integrand can be reduced using the trigonometric identity $1+\tan^2 x = \sec^2 x$ and the reduction formula:

$$\int \tan^n x dx = \int \tan^{n-2} x \tan^2 x dx = \int \tan^{n-2} x(\sec^2 x - 1)dx$$

$$= \frac{\tan^{n-1} x}{n-1} - \int \tan^{n-2} x dx.$$

4. Integrals of the form $\int \cot^n x dx$.

The power of the integrand can be reduced using the trigonometric identity $1+\cot^n x = \csc^n x$ and the reduction formula:

$$\int \cot^n x dx = \int \cot^{n-2} x \cot^2 x dx = \int \cot^{n-2} x(\csc^2 x - 1)dx = \frac{\cot^{n-1} x}{n-1} - \int \cot^{n-2} x dx.$$

5. Integrals of the form $\int \sec^n x dx$.

This type of integrals can be simplified with help of the reduction formula:

$$\int \sec^n x dx = \frac{\sec^{n-2} x \tan x}{n-1} + \frac{n-2}{n-1}\int \sec^{n-2} x dx.$$

6. Integrals of the form $\int \csc^n x dx$.

Similarly to the previous examples, this type of integrals can be simplified by the formula:

$$\int \csc^n x dx = \frac{\csc^{n-2} x \cot x}{n-1} + \frac{n-2}{n-1}\int \csc^{n-2} x dx.$$

7. Integrals of the form $\int \tan^m x \sec^n x dx$.

- If the power of the secant n is even, then using the identity $1+\tan^2 x=\sec^2 x$ the secant function is expressed as the tangent function. The factor $\sec^2 x$ is separated and used for transformation of the differential. As a result, the entire integral (including differential) is expressed in terms of the function tan x.

- If both the powers n and m are odd, then the factor sec x tan x, which is necessary to transform the differential, is separated. Then the entire integral is expressed in terms of sec x.

- If the power of the secant n is odd, and the power of the tangent m is even, then the tangent is expressed in terms of the secant using the identity $1+\tan^2 x=\sec^2 x$. After this substitution, you can calculate the integrals of the secant.

8. Integrals of the form $\int \cot^m x \csc^n x dx$.

- If the power of the cosecant n is even, then using the identity $1+\cot^2 x=\csc^2 x$ the cosecant function is expressed as the cotangent function. The factor $\csc^2 x$ is separated and used for transformation of the differential. As a result, the integrand and differential are expressed in terms of cot x.

- If both the powers n and m are odd, then the factor $\cot x \csc x$, which is necessary to transform the differential, is separated. Then the integral is expressed in terms of $\csc x$.

- If the power of the cosecant n is odd, and the power of the cotangent m is even, then the cotangent is expressed in terms of the cosecant using the identity $1+\cot^2 x=\csc^2 x$. After this substitution, you can find the integrals of the cosecant.

APPLICATIONS

It may not have direct applications in solving practical issues but used in the various field. For example, trigonometry is used in developing computer music: as you are familiar that sound travels in the form of waves and this wave pattern through a sine or cosine function for developing computer music. Here are a few applications where trigonometry and its functions are applicable.

Trigonometry in Aviation

Aviation technology has been evolved in many up-gradations in the last few years. It has taken into account the speed, direction, and distance as well as have to consider the speed and direction of the wind. The wind plays a vital role in when and how a flight will travel. This equation cab is solved by using trigonometry.

For example, if an aeroplane is travelling at 250 miles per hour, 55° of north of east and the wind blowing due to south at 19 miles per hour. This calculation will be solved using the trigonometry and find the third side of the triangle that will lead the aircraft in the right direction.

Trigonometry in Criminology

Trigonometry is even used in the investigation of a crime scene. The functions of trigonometry are helpful to calculate a trajectory of a projectile and to estimate the causes of a collision in a car accident. Further, it is used to identify how an object falls or in what angle the gun is shot.

Trigonometry in Marine Biology

Trigonometry is often used by marine biologists for measurements to figure out the depth of sunlight that affects algae to photosynthesis. Using the trigonometric function and mathematical models, marine biologists estimate the size of larger animals like whales and also understand their behaviours.

Trigonometry in Navigation

Trigonometry is used in navigating directions; it estimates in what direction to place the compass to get a straight direction. With the help of a compass and trigonometric functions in navigation, it will help to pinpoint a location and also to find distance as well to see the horizon.

Other uses of Trigonometry

- The calculus is based on trigonometry and algebra.

- The fundamental trigonometric functions like sine and cosine are used to describe the sound and light waves.
- Trigonometry is used in oceanography to calculate heights of waves and tides in oceans.
- It used in the creation of maps.
- It is used in satellite systems.

Heights and Distances

Trigonometry is used to in measuring the height of a building or a mountain. The distance of a building from the viewpoint and the elevation angle can easily determine the height of a building using the trigonometric functions.

Measuring Heights and Distances

To measure heights and distances of different objects, we use trigonometric ratios. For example, in figure, there is a guy looking at the top of the lamppost. AB is horizontal level. This level is the line parallel to ground passing through the observer's eyes. AC is known as the line of sight. ∠A is called angle of elevation. Similarly, in figure, PQ is the line of sight, PR is the horizontal level and ∠P is called angle of depression.

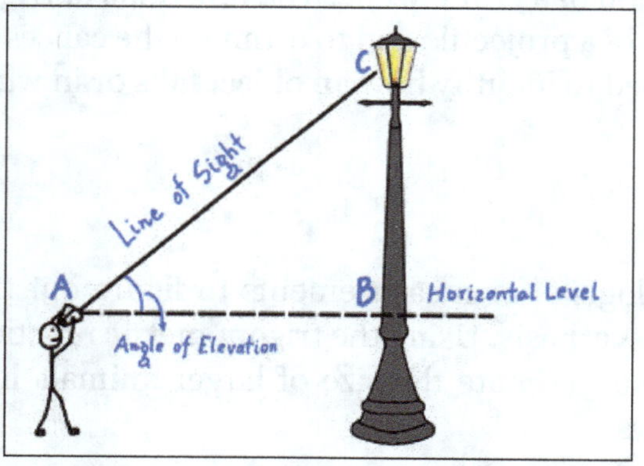

Angle of Elevation in Height And Distance.

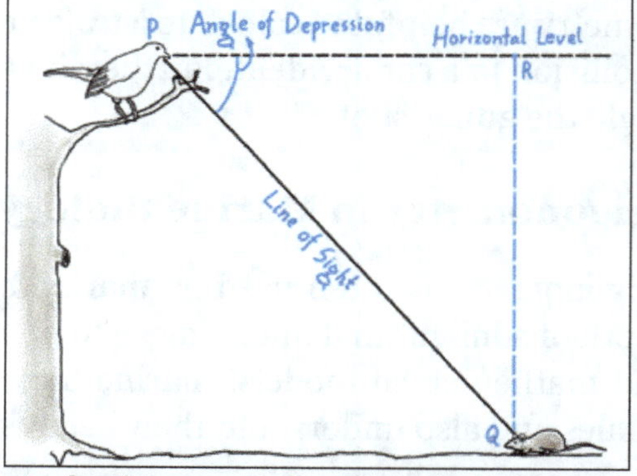

Angle of Depression in Height And Distance.

The angles of elevation and depression are usually measured by a device called Inclinometer or Clinometer.

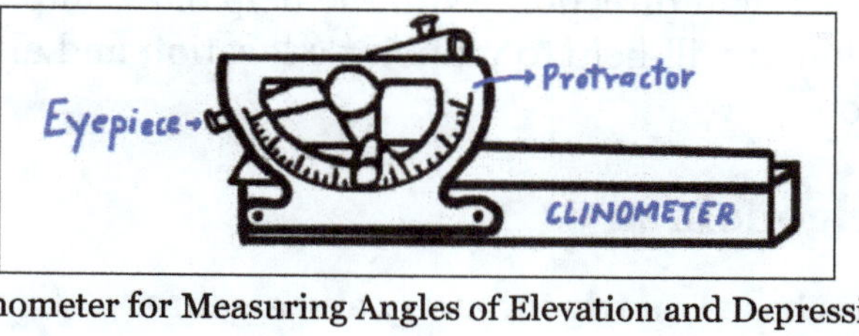

Inclinometer for Measuring Angles of Elevation and Depression.

To solve height and distance problems, trigonometric ratios for standard angles, following table should be remembered.

Table: Trigonometric Ratios of Standard Angles.

∠C	0°	30°	45°	60°	90°
sin C	0	½	$1/\sqrt{2}$	$\sqrt{3}/2$	1
cos C	1	$\sqrt{3}/2$	$1/\sqrt{2}$	½	0
tan C	0	$1/\sqrt{3}$	1	$\sqrt{3}$	Not Defined
csc C	Not Defined	2	$\sqrt{2}$	2/3	1
sec C	1	$2/\sqrt{3}$	$\sqrt{2}$	2	Not Defined
cot C	Not Defined	$\sqrt{3}$	1	$1/\sqrt{3}$	0

Example: An aeroplane is flying h meters above the ground. At a particular instant, the angle of elevation of the plane from the eyes of a boy sitting on the ground is 60°. After some time, the angle of elevation changed to 30°. Find the distance covered by the plane during that time assuming it travelled in a straight line.

Solution: The scenario explained in the question can be drawn as shown in the figure.

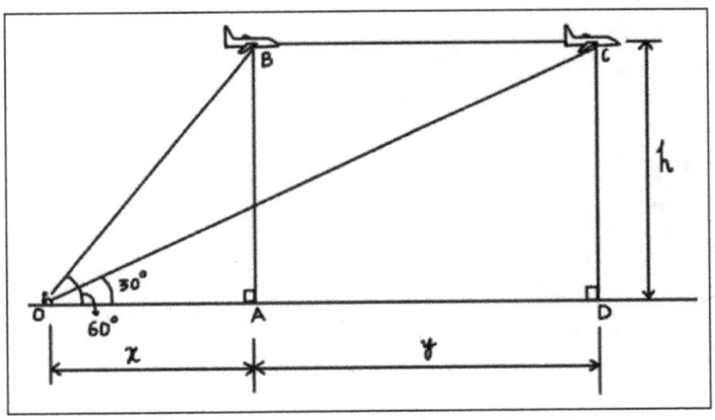

Height and Distance Example Question.

In Δ OAB,

$$\tan 60° = ABOA$$

$$\sqrt{3} = h/x$$

$$x = h/\sqrt{3}$$

In Δ OCD,

$$\tan 30° = CD/OD$$

$$1/\sqrt{3} = h/(x+y)$$

$$x+y = \sqrt{3}h$$

Distance travelled by plane = AD = y

$$(x+y)-x = \sqrt{3}h - h/\sqrt{3}$$

$$y = (2/\sqrt{3})h$$

So, if the aeroplane is flying h meters above the ground, it would travel for $(2/\sqrt{3})$ h meters as the angle of elevation changes from 60° to 30 °.

References

- Trigonometry: newworldencyclopedia.org, Retrieved 17 January, 2020
- Trigonometric-ratios, maths-introduction-to-trigonometry: toppr.com, Retrieved 19 February, 2020
- Applications-of-trigonometry, maths: byjus.com, Retrieved 23 June, 2020
- Height-and-distance, maths: byjus.com, Retrieved 16 May, 2020

2

Trigonometric Functions: An Integrated Study

Trigonometric functions are the angle functions between an angle of a right-angled triangle and ratios of two side lengths. It includes sine, cosine, tangent, cosecant, secant and cotangent. This chapter has been carefully written to provide an easy understanding of the trigonometric functions.

In Mathematics, trigonometric function is one of six functions (sine, cosine, tangent, cotangent, secant, and cosecant) that represent ratios of sides of right triangles. They are also known as the circular functions, since their values can be defined as ratios of the x and y coordinates of points on a circle of radius 1 that correspond to angles in standard positions. Such values have been tabulated and programmed into scientific calculators and computers. This allows trigonometry to be easily applied to surveying, engineering, and navigation problems in which one of a right triangle's acute angles and the length of a side are known and the lengths of the other sides are to be found. The fundamental trigonometric identity is $\sin^2\theta + \cos^2\theta = 1$, in which θ is an angle. Certain intrinsic qualities of the trigonometric functions make them useful in mathematical analysis. In particular, their derivatives form patterns useful for solving differential equations.

There are six main trigonometric functions:

- Sine (sin).
- Cosine (cos).
- Tangent (tan).
- Secant (sec).
- Cosecant (csc).
- Cotangent (cot).

These functions are used to relate the angles of a triangle with the sides of that triangle. Trigonometric functions are important when studying triangles and modeling periodic phenomena such as waves, sound, and light.

To define these functions for the angle theta, begin with a right triangle. Each function relates the angle to two sides of a right triangle. First, let's define the sides of the triangle.

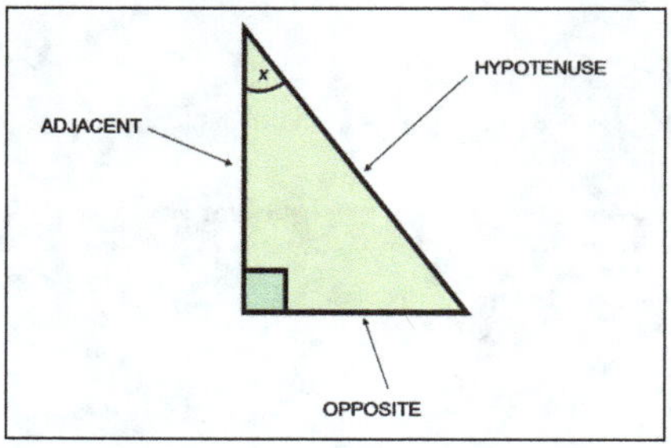

- The hypotenuse is the side opposite the right angle. The hypotenuse is always the longest side of a right triangle.
- The opposite side is the side opposite to the angle we are interested in, theta.
- The adjacent side is the side having both the angles of interest (angle theta and the right angle).

The relationship between the trigonometric functions and the sides of the triangle are as follows:

- Sine(theta) = opposite/hypotenuse.
- Cosecant(theta) = hypotenuse/opposite.
- Cosine(theta) = adjacent/hypotenuse.
- Secant(theta) = hypotenuse/adjacent.
- Tangent(theta) = opposite/adjacent.
- Cotangent(theta) = adjacent/opposite.

Real-world Examples

These trigonometric functions have practical applications in surveying, building, engineering, and even medicine. Here's one practical way to use these functions to solve a problem:

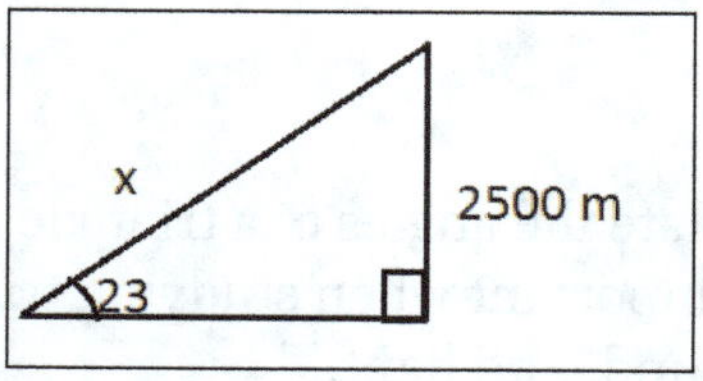

The angle of elevation of an airplane is 23 degrees, and its altitude is 2500 meters. How far away is it?

We are trying to solve this right triangle for the hypotenuse x. Since the side length we know is opposite the angle we know, we can use the sine function.

Sin(23) = 2500m/x

x = 6398.3 meters

SINE

In mathematics, the sine is a trigonometric function of an angle. The sine of an acute angle is defined in the context of a right triangle: for the specified angle, it is the ratio of the length of the side that is opposite that angle to the length of the longest side of the triangle (the hypotenuse).

More generally, the definition of sine (and other trigonometric functions) can be extended to any real value in terms of the length of a certain line segment in a unit circle. More modern definitions express the sine as an infinite series or as the solution of certain differential equations, allowing their extension to arbitrary positive and negative values and even to complex numbers.

The sine function is commonly used to model periodic phenomena such as sound and light waves, the position and velocity of harmonic oscillators, sunlight intensity and day length, and average temperature variations throughout the year.

Right-angled Triangle Definition

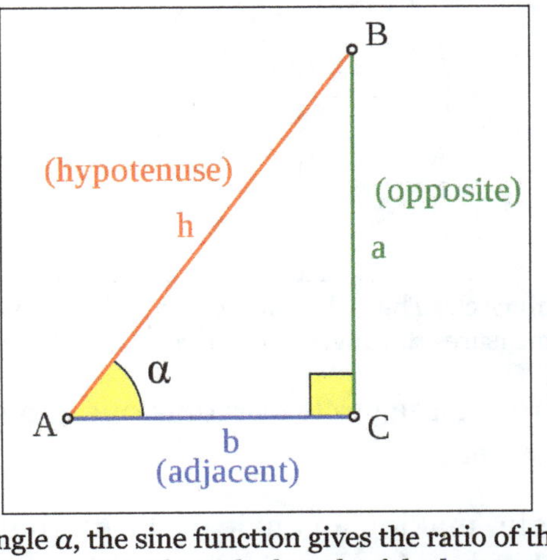

For the angle α, the sine function gives the ratio of the length of the opposite side to the length of the hypotenuse.

To define the sine function of an acute angle α, start with a right triangle that contains an angle of measure α; in the accompanying figure, angle A in triangle ABC is the angle of interest. The three sides of the triangle are named as follows:

- The opposite side is the side opposite to the angle of interest, in this case side a.
- The hypotenuse is the side opposite the right angle, in this case side h. The hypotenuse is always the longest side of a right-angled triangle.
- The adjacent side is the remaining side, in this case side b. It forms a side of (is adjacent to) both the angle of interest (angle A) and the right angle.

Once such a triangle is chosen, the sine of the angle is equal to the length of the opposite side divided by the length of the hypotenuse, or:

$$\sin(\alpha) = \frac{\text{opposite}}{\text{hypotenuse}}$$

The other trigonometric functions of the angle can be defined similarly; for example, the cosine of the angle is the ratio between the adjacent side and the hypotenuse, while the tangent gives the ratio between the opposite and adjacent sides.

As stated, the value $\sin(\alpha)$ appears to depend on the choice of right triangle containing an angle of measure α. However, this is not the case: all such triangles are similar, and so the ratio is the same for each of them.

Unit Circle

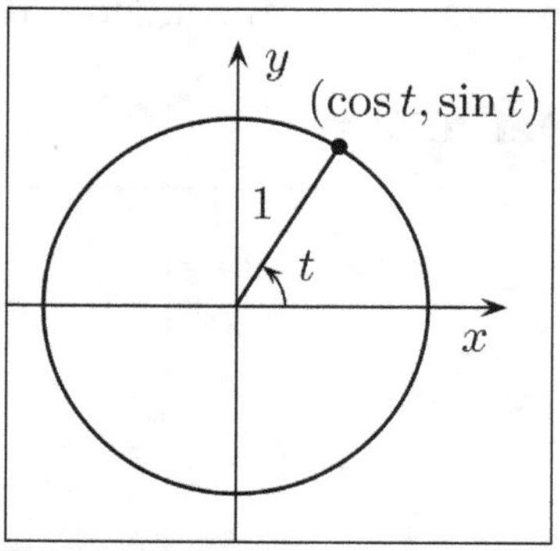

Unit circle: The radius has length 1. The variable t measures the angle referred to as θ in the text.

In trigonometry, a unit circle is the circle of radius one centered at the origin (0, 0) in the Cartesian coordinate system.

Let a line through the origin, making an angle of θ with the positive half of the x-axis, intersect the unit circle. The x- and y-coordinates of this point of intersection are equal to cos(θ) and sin(θ), respectively. The point's distance from the origin is always 1.

Unlike the definitions with the right triangle or slope, the angle can be extended to the full set of real arguments by using the unit circle. This can also be achieved by requiring certain symmetries and that sine be a periodic function.

Identities

Exact identities (using radians):

These apply for all values of θ.

$$\sin(\theta) = \cos\left(\frac{\pi}{2} - \theta\right) = \frac{1}{\csc(\theta)}$$

Reciprocal

The reciprocal of sine is cosecant, i.e., the reciprocal of sin(A) is csc(A), or cosec(A). Cosecant gives the ratio of the length of the hypotenuse to the length of the opposite side:

$$\csc(A) = \frac{1}{\sin(A)} = \frac{\text{hypotenuse}}{\text{opposite}} = \frac{h}{a}.$$

Inverse

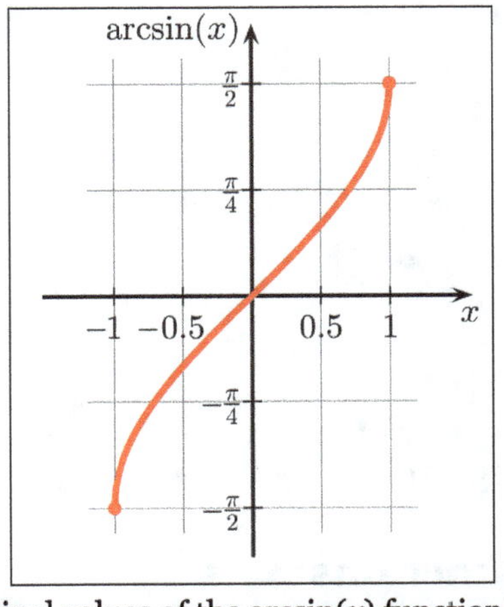

The usual principal values of the arcsin(*x*) function graphed on the cartesian plane. Arcsin is the inverse of sin.

The inverse function of sine is arcsine (arcsin or asin) or inverse sine ($\sin^{-1}$). As sine is non-injective, it is not an exact inverse function but a partial inverse function. For example, sin(0) = 0, but also sin(π) = 0, sin(2π) = 0 etc. It follows that the arcsine function is multivalued: arcsin(0) = 0, but also arcsin(0) = π, arcsin(0) = 2π, etc. When only one value is desired, the function may be restricted to its principal branch. With this restriction, for each x in the domain the expression arcsin(x) will evaluate only to a single value, called its principal value.

$$\theta = \arcsin\left(\frac{\text{opposite}}{\text{hypotenuse}}\right) = \sin^{-1}\left(\frac{a}{h}\right).$$

k is some integer:

$$\sin(y) = x \Leftrightarrow y = \arcsin(x) + 2\pi k, \text{or}$$
$$y = \pi - \arcsin(x) + 2\pi k$$

Or in one equation:

$$\sin(y) = x \Leftrightarrow y = (-1)^k \arcsin(x) + \pi k$$

Arcsin satisfies:

$$\sin(\arcsin(x)) = x$$

and

$$\arcsin(\sin(\theta)) = \theta \quad \text{for } -\frac{\pi}{2} \le \theta \le \frac{\pi}{2}.$$

Calculus

For the sine function:

$$f(x) = \sin(x)$$

The derivative is:

$$f'(x) = \cos(x)$$

The antiderivative is:

$$\int f(x)dx = -\cos(x) + C$$

C denotes the constant of integration.

Other Trigonometric Functions

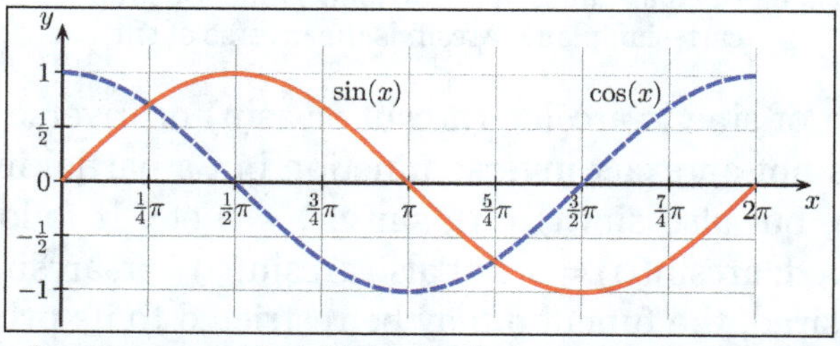

The sine and cosine functions are related in multiple ways. The two functions are out of phase by 90°: $\sin(\pi/2 - x) = \cos(x)$ for all angles x. Also, the derivative of the function sin(x) is cos(x).

It is possible to express any trigonometric function in terms of any other (up to a plus or minus sign, or using the sign function).

Sine in terms of the other common trigonometric functions:

Trigonometric functions:

	fθ	Using plus/minus (±)					Using sign function (sgn)
		fθ =	± per Quadrant				fθ =
			I	II	III	IV	
cos	sin(θ)	$=\pm\sqrt{1-\cos^2(\theta)}$	+	+	−	−	$=\operatorname{sgn}\left(\cos\left(\theta-\frac{\pi}{2}\right)\right)\sqrt{1-\cos^2(\theta)}$
cos	cos(θ)	$=\pm\sqrt{1-\sin^2(\theta)}$	+	−	−	+	$=\operatorname{sgn}\left(\sin\left(\theta+\frac{\pi}{2}\right)\right)\sqrt{1-\sin^2(\theta)}$
cot	sin(θ)	$=\pm\frac{1}{\sqrt{1+\cot^2(\theta)}}$	+	+	−	−	$=\operatorname{sgn}\left(\cot\left(\frac{\theta}{2}\right)\right)\frac{1}{\sqrt{1+\cot^2(\theta)}}$
cot	cot(θ)	$=\pm\frac{\sqrt{1-\sin^2(\theta)}}{\sin(\theta)}$	+	−	−	+	$=\operatorname{sgn}\left(\sin\left(\theta+\frac{\pi}{2}\right)\right)\frac{\sqrt{1-\sin^2(\theta)}}{\sin(\theta)}$
tan	sin(θ)	$=\pm\frac{\tan(\theta)}{\sqrt{1+\tan^2(\theta)}}$	+	−	−	+	$=\operatorname{sgn}\left(\tan\left(\frac{2\theta+\pi}{4}\right)\right)\frac{\tan(\theta)}{\sqrt{1+\tan^2(\theta)}}$
tan	tan(θ)	$=\pm\frac{\sin(\theta)}{\sqrt{1-\sin^2(\theta)}}$	+	−	−	+	$=\operatorname{sgn}\left(\sin\left(\theta+\frac{\pi}{2}\right)\right)\frac{\sin(\theta)}{\sqrt{1-\sin^2(\theta)}}$
sec	sin(θ)	$=\pm\frac{\sqrt{\sec^2(\theta)-1}}{\sec(\theta)}$	+	−	+	−	$=\operatorname{sgn}\left(\sec\left(\frac{4\theta-\pi}{2}\right)\right)\frac{\sqrt{\sec^2(\theta)-1}}{\sec(\theta)}$
sec	sec(θ)	$=\pm\frac{1}{\sqrt{1-\sin^2(\theta)}}$	+	−	−	+	$=\operatorname{sgn}\left(\sin\left(\theta+\frac{\pi}{2}\right)\right)\frac{1}{\sqrt{1-\sin^2(\theta)}}$

For all equations which use plus/minus (±), the result is positive for angles in the first quadrant.

The basic relationship between the sine and the cosine can also be expressed as the Pythagorean trigonometric identity:

$$\cos^2(\theta)+\sin^2(\theta)=1$$

where $\sin^2(x)$ means $(\sin(x))^2$.

Sine Squared Function

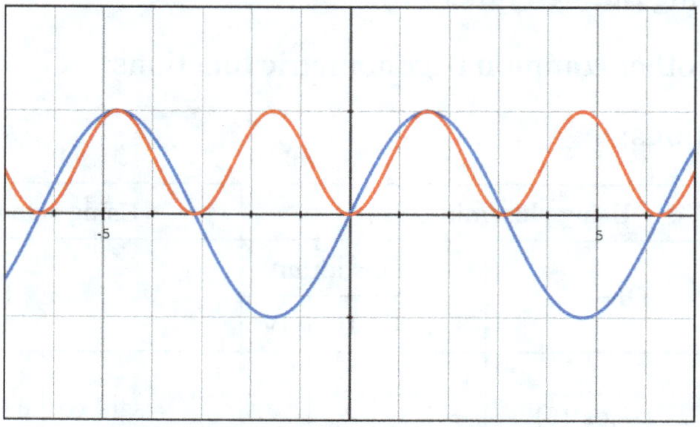

Sine function in blue and sine squared function in red. The Y axis is in radians.

The graph shows both the sine function and the sine squared function, with the sine in blue and sine squared in red. Both graphs have the same shape, but with different ranges of values, and different periods. Sine squared has only positive values, but twice the number of periods.

The sine squared function can be expressed as a modified sine wave from the Pythagorean identity and power reduction by the cosine double-angle formula:

$$\sin^2(\theta) = \frac{1-\sin(2\theta+\frac{\pi}{2})}{2}$$

Properties Relating to the Quadrants

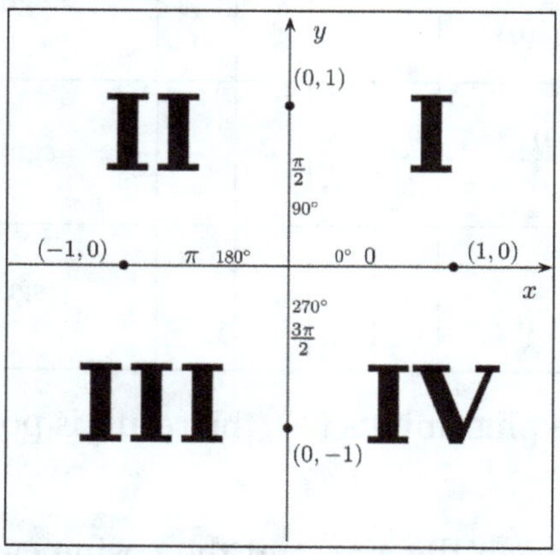

The four quadrants of a Cartesian coordinate system.

The table below displays many of the key properties of the sine function (sign, monotonicity, convexity), arranged by the quadrant of the argument. For arguments outside those in the table, one may compute the corresponding information by using the periodicity $\sin(\alpha+360^\circ)=\sin(\alpha)$ of the sine function.

Quadrant	Degrees	Radians	Value	Sign	Monotony	Convexity
1st Quadrant	$0^\circ < x < 90^\circ$	$0 < x < \frac{\pi}{2}$	$0 < \sin(x) < 1$	+	Increasing	Concave
2nd Quadrant	$90^\circ < x < 180^\circ$	$\frac{\pi}{2} < x < \pi$	$0 < \sin(x) < 1$	+	Decreasing	Concave
3rd Quadrant	$180^\circ < x < 270^\circ$	$\pi < x < \frac{3\pi}{2}$	$-1 < \sin(x) < 0$	-	Decreasing	Convex
4th Quadrant	$270^\circ < x < 360^\circ$	$\frac{3\pi}{2} < x < 2\pi$	$-1 < \sin(x) < 0$	-	Increasing	Convex

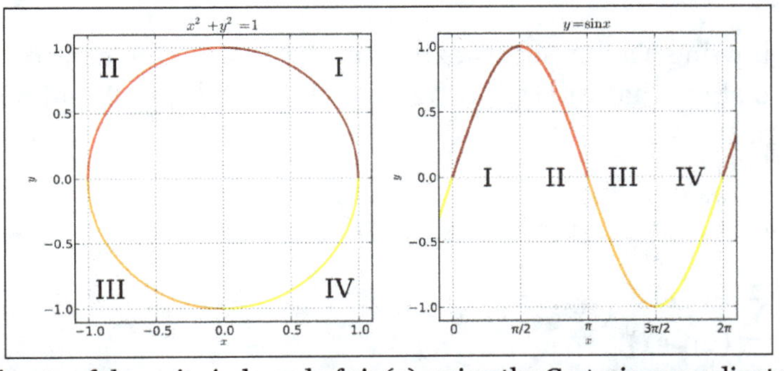

The quadrants of the unit circle and of sin(x), using the Cartesian coordinate system.

The following table gives basic information at the boundary of the quadrants:

Degrees	Radians	sin(x)	Point type
0°	0	0	Root, Inflection
90°	$\frac{\pi}{2}$	1	Maximum
180°	π	0	Root, Inflection
270°	$\frac{3\pi}{2}$	-1	Minimum

Series

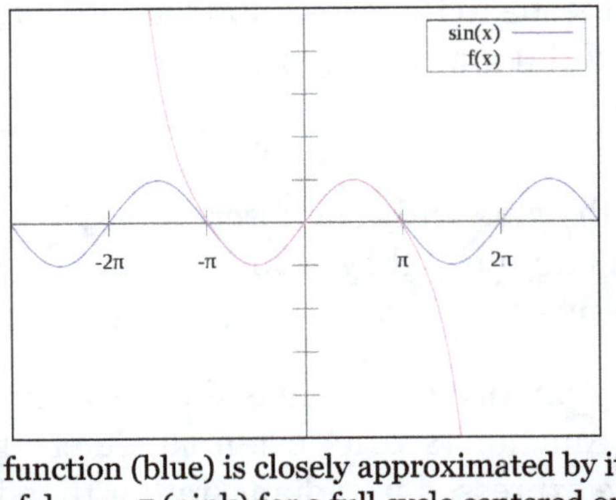

The sine function (blue) is closely approximated by its Taylor polynomial of degree 7 (pink) for a full cycle centered on the origin.

Using only geometry and properties of limits, it can be shown that the derivative of sine is cosine, and that the derivative of cosine is the negative of sine.

Using the reflection from the calculated geometric derivation of the sine is with the (4n+k)-th derivative at the point 0:

$$\sin^{(4n+k)}(0) = \begin{cases} 0 & \text{when } k = 0 \\ 1 & \text{when } k = 1 \\ 0 & \text{when } k = 2 \\ -1 & \text{when } k = 3 \end{cases}$$

This gives the following Taylor series expansion at x = 0. One can then use the theory of Taylor series to show that the following identities hold for all real numbers x (where x is the angle in radians):

$$\begin{aligned} \sin(x) &= x - \frac{x^3}{3!} + \frac{x^5}{5!} - \frac{x^7}{7!} + \cdots \\ &= \sum_{n=0}^{\infty} \frac{(-1)^n}{(2n+1)!} x^{2n+1} \end{aligned}$$

If x were expressed in degrees then the series would contain factors involving powers of $\pi/180$: if x is the number of degrees, the number of radians is $y = \pi x/180$, so:

$$\begin{aligned} \sin(x_{deg}) &= \sin(y_{rad}) \\ &= \frac{\pi}{180}x - \left(\frac{\pi}{180}\right)^3 \frac{x^3}{3!} + \left(\frac{\pi}{180}\right)^5 \frac{x^5}{5!} - \left(\frac{\pi}{180}\right)^7 \frac{x^7}{7!} + \cdots \end{aligned}$$

The series formulas for the sine and cosine are uniquely determined, up to the choice of unit for angles, by the requirements that,

$$\sin(0) = 0 \text{ and } \sin(2x) = 2\sin(x)\cos(x)$$

$$\cos^2(x) + \sin^2(x) = 1 \text{ and } \cos(2x) = \cos^2(x) - \sin^2(x)$$

The radian is the unit that leads to the expansion with leading coefficient 1 for the sine and is determined by the additional requirement that,

$$\sin(x) \approx x \text{ when } x \approx 0.$$

The coefficients for both the sine and cosine series may therefore be derived by substituting their expansions into the pythagorean and double angle identities, taking the leading coefficient for the sine to be 1, and matching the remaining coefficients.

In general, mathematically important relationships between the sine and cosine functions and the exponential function (see, for example, Euler's formula) are substantially simplified when angles are expressed in radians, rather than in degrees, grads or other units. Therefore, in most branches of mathematics beyond practical geometry, angles are generally assumed to be expressed in radians.

A similar series is Gregory's series for arctan, which is obtained by omitting the factorials in the denominator.

Continued Fraction

The sine function can also be represented as a generalized continued fraction:

$$\sin(x) = \cfrac{x}{1 + \cfrac{x^2}{2\cdot 3 - x^2 + \cfrac{2\cdot 3x^2}{4\cdot 5 - x^2 + \cfrac{4\cdot 5x^2}{6\cdot 7 - x^2 + \ddots}}}}$$

The continued fraction representation can be derived from Euler's continued fraction formula and expresses the real number values, both rational and irrational, of the sine function.

Fixed Point

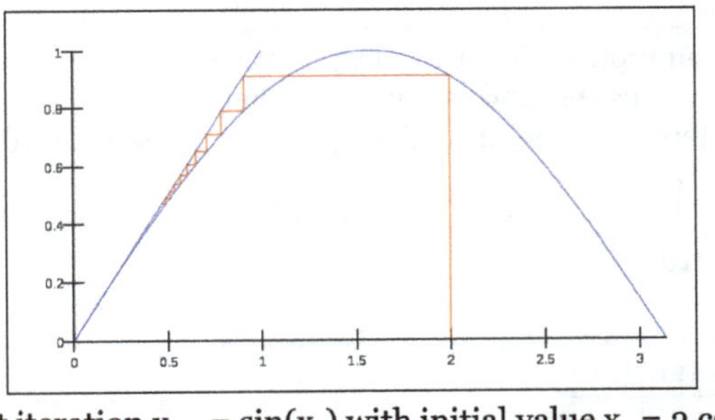

The fixed point iteration $x_{n+1} = \sin(x_n)$ with initial value $x_0 = 2$ converges to 0.

Zero is the only real fixed point of the sine function; in other words the only intersection of the sine function and the identity function is $\sin(0) = 0$.

Arc Length

The arc length of the sine curve between a and b is, $\int_a^b \sqrt{1+\cos^2(x)}\,dx$. This integral is an elliptic integral of the second kind.

The arc length for a full period is $\frac{4\sqrt{2\pi^3}}{\Gamma(1/4)^2} + \frac{\Gamma(1/4)^2}{\sqrt{2\pi}} = 7.640395578\ldots$ where Γ is the gamma function.

The arc length of the sine curve from 0 to x is the above number divided by 2π times x, plus a correction that varies periodically in x with period π. The Fourier series for this correction can be written in closed form using special functions, but it is perhaps more instructive to write the decimal approximations of the Fourier coefficients. The sine curve arc length from 0 to x is,

$$1.21600672x + 0.10317093\sin(2x) - 0.00220445\sin(4x) + 0.00012584\sin(6x)$$
$$-0.00001011\sin(8x) + \cdots$$

Special Values

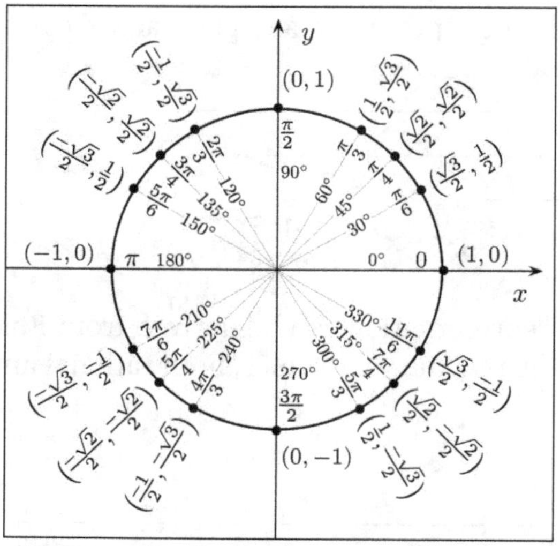

Some common angles (θ) shown on the unit circle. The angles are given in degrees and radians, together with the corresponding intersection point on the unit circle, (cos(θ), sin(θ)).

For certain integral numbers *x* of degrees, the value of sin(*x*) is particularly simple. A table of some of these values is given below:

<table>
<tr><th colspan="4">x (angle)</th><th colspan="2">sin(x)</th></tr>
<tr><th>Degrees</th><th>Radians</th><th>Gradians</th><th>Turns</th><th>Exact</th><th>Decimal</th></tr>
<tr><td>$0°$</td><td>0</td><td>0^g</td><td>0</td><td rowspan="2">0</td><td rowspan="2">0</td></tr>
<tr><td>$180°$</td><td>π</td><td>200^g</td><td>$\frac{1}{2}$</td></tr>
<tr><td>$15°$</td><td>$\frac{1}{12}\pi$</td><td>$16\frac{2}{3}g$</td><td>$\frac{1}{24}$</td><td rowspan="2">$\frac{\sqrt{6}-\sqrt{2}}{2}$</td><td rowspan="2">0.258819045102521</td></tr>
<tr><td>$165°$</td><td>$\frac{11}{12}\pi$</td><td>$183\frac{1}{3}g$</td><td>$\frac{11}{24}$</td></tr>
<tr><td>$30°$</td><td>$\frac{1}{6}\pi$</td><td>$33\frac{1}{3}g$</td><td>$\frac{1}{12}$</td><td rowspan="2">$\frac{1}{2}$</td><td rowspan="2">0.5</td></tr>
<tr><td>$150°$</td><td>$\frac{5}{6}\pi$</td><td>$166\frac{2}{3}g$</td><td>$\frac{5}{12}$</td></tr>
</table>

<table>
<tr><td>45°</td><td>$\frac{1}{4}\pi$</td><td>50^g</td><td>$\frac{1}{8}$</td><td rowspan="2">$\frac{\sqrt{2}}{2}$</td><td rowspan="2">0.707106781186548</td></tr>
<tr><td>135°</td><td>$\frac{3}{4}\pi$</td><td>150^g</td><td>$\frac{3}{8}$</td></tr>
<tr><td>60°</td><td>$\frac{1}{3}\pi$</td><td>$66\frac{2}{3}g$</td><td>$\frac{1}{6}$</td><td rowspan="2">$\frac{\sqrt{3}}{2}$</td><td rowspan="2">0.866025403784439</td></tr>
<tr><td>120°</td><td>$\frac{2}{3}\pi$</td><td>$133\frac{1}{3}g$</td><td>$\frac{1}{3}$</td></tr>
<tr><td>75°</td><td>$\frac{5}{12}\pi$</td><td>$83\frac{1}{3}g$</td><td>$\frac{5}{24}$</td><td rowspan="2">$\frac{\sqrt{6}+\sqrt{2}}{4}$</td><td rowspan="2">0.965925826289068</td></tr>
<tr><td>105°</td><td>$\frac{7}{12}\pi$</td><td>$116\frac{2}{3}g$</td><td>$\frac{7}{24}$</td></tr>
<tr><td>90°</td><td>$\frac{1}{2}\pi$</td><td>100^g</td><td>$\frac{1}{4}$</td><td>1</td><td>1</td></tr>
</table>

90 degree increments:

x in degrees	0°	90°	180°	270°	360°
x in radians	0	π/2	π	3π/2	2π
x in gons	0	100^g	200^g	300^g	400^g
x in turns	0	1/4	1/2	3/4	1
sin x	0	1	0	-1	0

Other values not listed above:

$$\sin\left(\frac{\pi}{60}\right) = \sin(3^\circ) = \frac{(2-\sqrt{12})\sqrt{5+\sqrt{5}}+(\sqrt{10}-\sqrt{2})(\sqrt{3}+1)}{16}$$

$$\sin\left(\frac{\pi}{30}\right) = \sin(6^\circ) = \frac{\sqrt{30-\sqrt{180}}-\sqrt{5}-1}{8}$$

$$\sin\left(\frac{\pi}{20}\right) = \sin(9^\circ) = \frac{\sqrt{10}+\sqrt{2}-\sqrt{20-\sqrt{80}}}{8}$$

$$\sin\left(\frac{\pi}{15}\right) = \sin(12^\circ) = \frac{\sqrt{10+\sqrt{20}}+\sqrt{3}-\sqrt{15}}{8}$$

$$\sin\left(\frac{\pi}{10}\right) = \sin(18^\circ) = \frac{\sqrt{5}-1}{4} = \tfrac{1}{2}\varphi^{-1}$$

$$\sin\left(\frac{7\pi}{60}\right)=\sin(21^\circ)=\frac{(2+\sqrt{12})\sqrt{5-\sqrt{5}}-(\sqrt{10}+\sqrt{2})(\sqrt{3}-1)}{16}$$

$$\sin\left(\frac{\pi}{8}\right)=\sin(22.5^\circ)=\frac{\sqrt{2-\sqrt{2}}}{2}$$

$$\sin\left(\frac{2\pi}{15}\right)=\sin(24^\circ)=\frac{\sqrt{3}+\sqrt{15}-\sqrt{10-\sqrt{20}}}{8}$$

$$\sin\left(\frac{3\pi}{20}\right)=\sin(27^\circ)=\frac{\sqrt{20+\sqrt{80}}-\sqrt{10}+\sqrt{2}}{8}$$

$$\sin\left(\frac{11\pi}{60}\right)=\sin(33^\circ)=\frac{(\sqrt{12}-2)\sqrt{5+\sqrt{5}}+(\sqrt{10}-\sqrt{2})(\sqrt{3}+1)}{16}$$

$$\sin\left(\frac{\pi}{5}\right)=\sin(36^\circ)=\frac{\sqrt{10-\sqrt{20}}}{4}$$

$$\sin\left(\frac{13\pi}{60}\right)=\sin(39^\circ)=\frac{(2-\sqrt{12})\sqrt{5-\sqrt{5}}+(\sqrt{10}+\sqrt{2})(\sqrt{3}+1)}{16}$$

$$\sin\left(\frac{7\pi}{30}\right)=\sin(42^\circ)=\frac{\sqrt{30+\sqrt{180}}-\sqrt{5}+1}{8}$$

Relationship to Complex Numbers

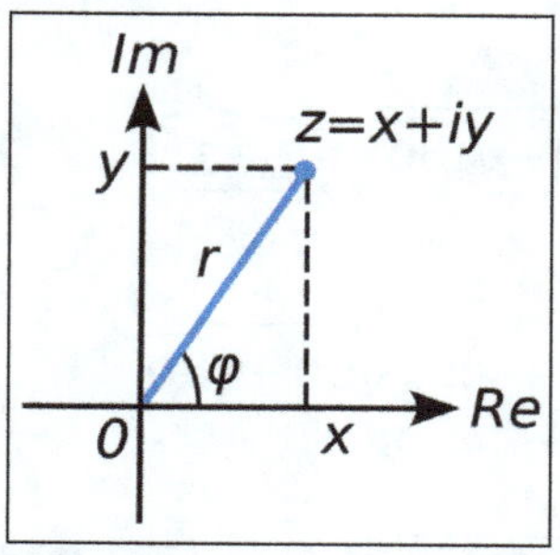

An illustration of the complex plane. The imaginary numbers are on the vertical coordinate axis.

Sine is used to determine the imaginary part of a complex number given in polar coordinates (r, φ):

$$z=r(\cos(\varphi)+i\sin(\varphi))$$

the imaginary part is:

$$\mathrm{Im}(z) = r\sin(\varphi)$$

r and φ represent the magnitude and angle of the complex number respectively. i is the imaginary unit. z is a complex number.

Although dealing with complex numbers, sine's parameter in this usage is still a real number. Sine can also take a complex number as an argument.

Sine with a Complex Argument

Domain coloring of sin(z) in the complex plane. Brightness indicates absolute magnitude, saturation represents complex argument.

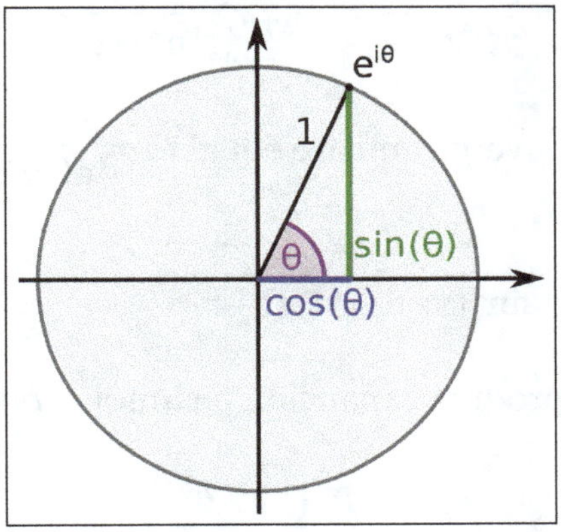

$\sin(\theta)$ is the imaginary part of $e^{i\theta}$.

The definition of the sine function for complex arguments z:

$$\begin{aligned}\sin(z) &= \sum_{n=0}^{\infty} \frac{(-1)^n}{(2n+1)!} z^{2n+1} \\ &= \frac{e^{iz} - e^{-iz}}{2i} \\ &= \frac{\sinh(iz)}{i}\end{aligned}$$

where $i^2 = -1$, and sinh is hyperbolic sine. This is an entire function. Also, for purely real x,

$$\sin(x) = \mathrm{Im}(e^{ix}).$$

For purely imaginary numbers:

$$\sin(iy) = i\sinh(y).$$

It is also sometimes useful to express the complex sine function in terms of the real and imaginary parts of its argument:

$$\begin{aligned}\sin(x+iy) &= \sin(x)\cos(iy)+\cos(x)\sin(iy)\\ &= \sin(x)\cosh(y)+i\cos(x)\sinh(y).\end{aligned}$$

Partial Fraction and Product Expansions of Complex Sine

Using the partial fraction expansion technique in complex analysis, one can find that the infinite series,

$$\sum_{n=-\infty}^{\infty}\frac{(-1)^n}{z-n}=\frac{1}{z}-2z\sum_{n=1}^{\infty}\frac{(-1)^n}{n^2-z^2}$$

both converge and are equal to $\frac{\pi}{\sin(\pi z)}$. Similarly, one can show that,

$$\frac{\pi^2}{\sin^2(\pi z)}=\sum_{n=-\infty}^{\infty}\frac{1}{(z-n)^2}.$$

Using product expansion technique, one can derive,

$$\sin(\pi z)=\pi z\prod_{n=1}^{\infty}\left(1-\frac{z^2}{n^2}\right).$$

Alternatively, the infinite product for the sine can be proved using complex Fourier series.

Proof of the Infinite Product for the Sine

Using complex Fourier series, the function $\cos(zx)$ can be decomposed as,

$$\cos(zx)=\frac{z\sin(\pi z)}{\pi}\sum_{n=-\infty}^{\infty}\frac{(-1)^n e^{inx}}{z^2-n^2}, z\in\mathbb{C}\setminus\{\mathbb{Z}\}, x\in[-\pi,\pi].$$

Setting $x=\pi$ yields

$$\cos(\pi z)=\frac{z\sin(\pi z)}{\pi}\sum_{n=-\infty}^{\infty}\frac{1}{z^2-n^2}=\frac{z\sin(\pi z)}{\pi}\left(\frac{1}{z^2}+2\sum_{n=1}^{\infty}\frac{1}{z^2-n^2}\right).$$

Therefore we get,

$$\pi\cot(\pi z)=\frac{1}{z}+2\sum_{n=1}^{\infty}\frac{z}{z^2-n^2}.$$

The function $\pi\cot(\pi z)$ is the derivative of $\ln(\sin(\pi z))+C_0$. Furthermore, if $\frac{df}{dz}=\frac{z}{z^2-n^2}$, then the function f such that the emerged series converges is $f=\frac{1}{2}\ln(1-z^2/n^2)+C_1$ which can be proved using the Weierstrass M-test. It follows that,

$$\ln(\sin(\pi z))=\ln(z)+\sum_{n=1}^{\infty}\ln\left(1-\frac{z^2}{n^2}\right)+C.$$

Exponentiating gives,

$$\sin(\pi z)=ze^{C}\prod_{n=1}^{\infty}\left(1-\frac{z^2}{n^2}\right).$$

Since $\lim_{z\to 0}\frac{\sin(\pi z)}{z}=\pi$ and $\lim_{z\to 0}\prod_{n=1}^{\infty}\left(1-\frac{z^2}{n^2}\right)=1$, we have $e^C=\pi$.

Hence,

$$\sin(\pi z)=\pi z\prod_{n=1}^{\infty}\left(1-\frac{z^2}{n^2}\right).$$

Let $a_n(z)=-\frac{z^2}{n^2}$. Since $\sum_{n=1}^{\infty}|a_n(z)|$ converges uniformly on any closed disk, $\prod_{n=1}^{\infty}(1+a_n(z))$ converges uniformly on any closed disk as well. Therefore the infinite product for the sine is valid for all $z\in\mathbb{C}$, which completes the proof.

Usage of Complex Sine

sin(z) is found in the functional equation for the Gamma function,

$$\Gamma(s)\Gamma(1-s)=\frac{\pi}{\sin(\pi s)},$$

which in turn is found in the functional equation for the Riemann zeta-function,

$$\zeta(s)=2(2\pi)^{s-1}\Gamma(1-s)\sin(\pi s/2)\zeta(1-s).$$

As a holomorphic function, sin z is a 2D solution of Laplace's equation:

$$\Delta u(x_1,x_2)=0.$$

The complex sine function is also related to the level curves of pendulums.

Complex Graphs

Sine function in the complex plane		
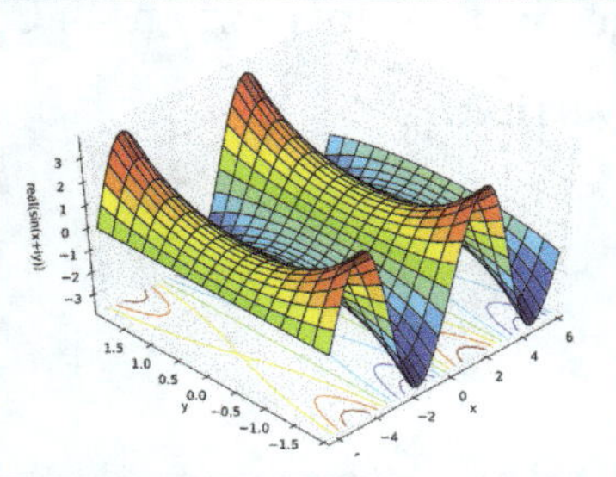	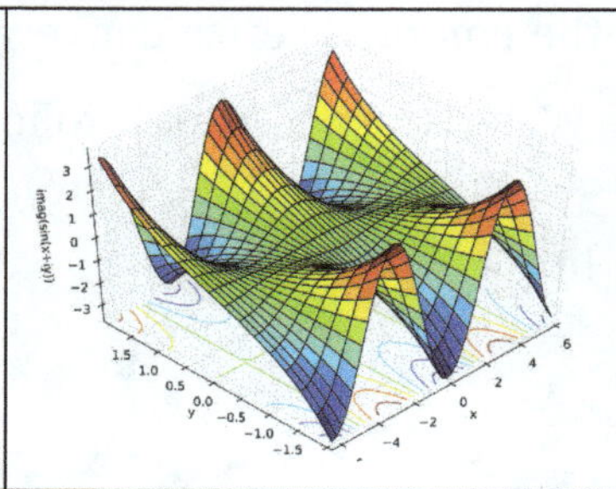	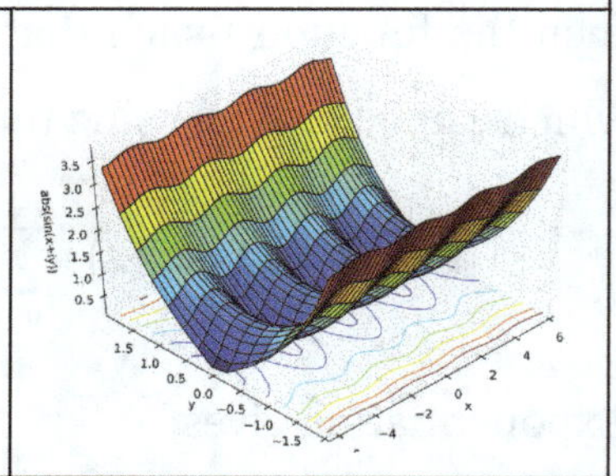
Real component	Imaginary component	Magnitude

Arcsine function in the complex plane		
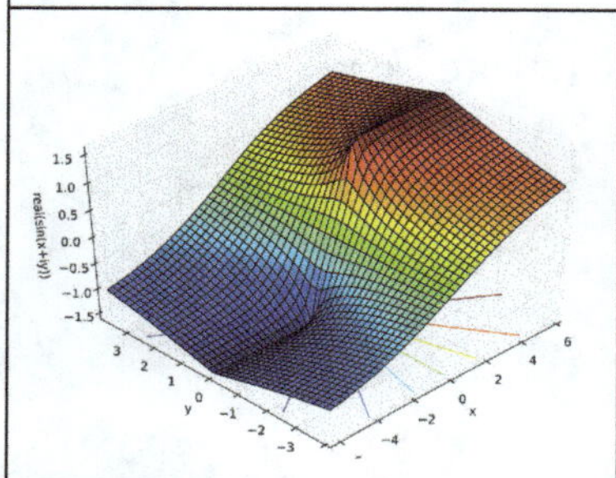	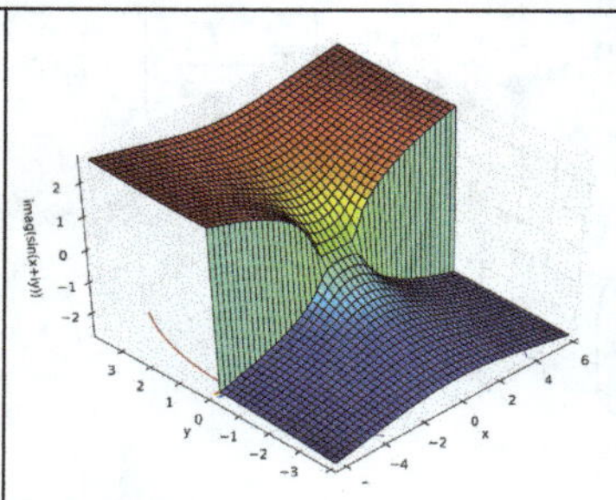	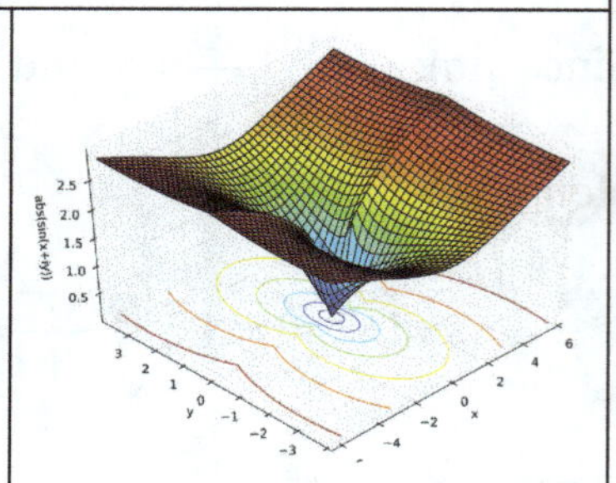
Real component	Imaginary component	Magnitude

COSINE

The cosine function is one of the oldest mathematical functions.

The cosine function is one of the oldest mathematical functions. It was first used in ancient Egypt in the book of Ahmes. Much later F. Viète evaluated some values of $\cos(nz)$, E. Gunter (introduced the notation "Cosi" and the word "cosinus" (replacing "complementi sinus"), and I. Newton found the series expansion for $\cos(z)$.

The classical definition of the cosine function for real arguments is: "the cosine of an angle α in a right-angle triangle is the ratio of the length of the adjacent leg to the length of the hypotenuse." This description of $\cos(\alpha)$, is valid for $0 < \alpha < \pi/2$ when the triangle is nondegenerate. This approach to the cosine can be expanded to arbitrary real values of α if consideration is given to the arbitrary point (x,y) in the x,y,-Cartesian plane and $\cos(\alpha)$ is defined as the ratio $x/\left(x^2+y^2\right)^{1/2}$, assuming that α is the value of the angle between the positive direction of the x-axis and the direction from the origin to the point.

The following formula can also be used as a definition of the cosine function:

$$\cos(z) = 1 - \frac{z^2}{2} + \frac{z^4}{24} - \ldots = \sum_{k=0}^{\infty} \frac{(-1)^k z^{2k}}{(2k)!}$$

This series converges for all finite numbers z.

The simplest way to understand the cosine function is to use the unit circle. For a given angle measure θ, draw a unit circle on the coordinate plane and draw the angle centered at the origin, with one side as the positive x -axis. The x-coordinate of the point where the other side of the angle intersects the circle is cos(θ), and the y-coordinate is sin(θ).

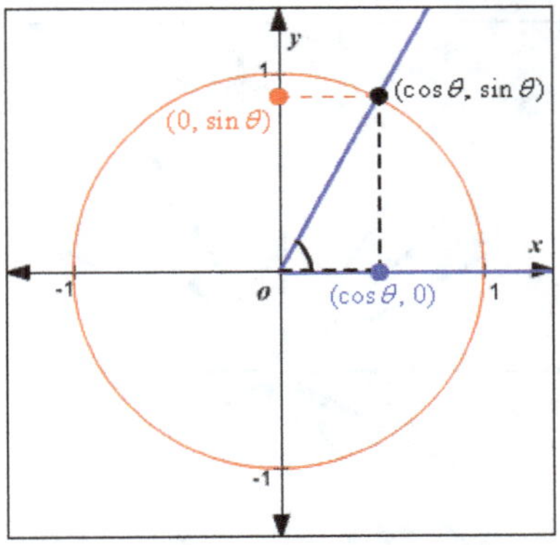

There are a few cosine values that should be memorized, based on 30°–60°–90° triangles and 45°–45°–90° triangles.

Once you know these values, you can derive many other values for the cosine function. Remember that cos\theta; is positive in quadrants I and IV and negative in quadrants II and III.

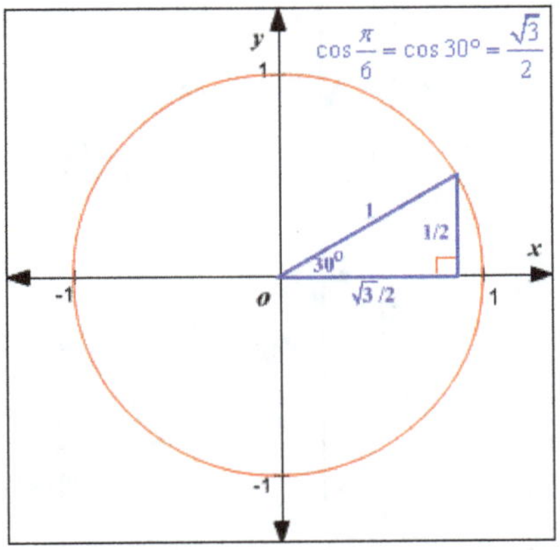

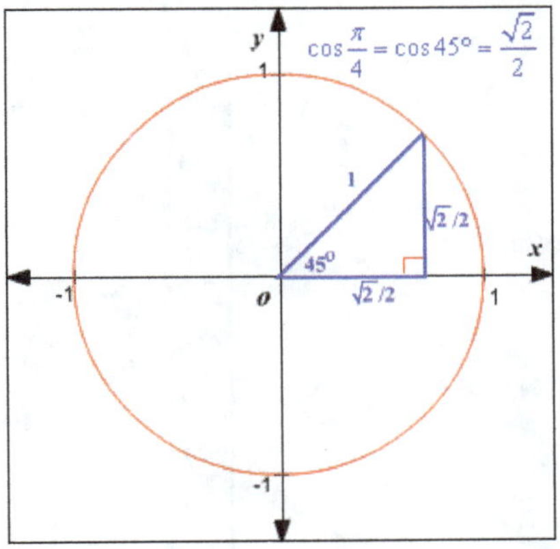

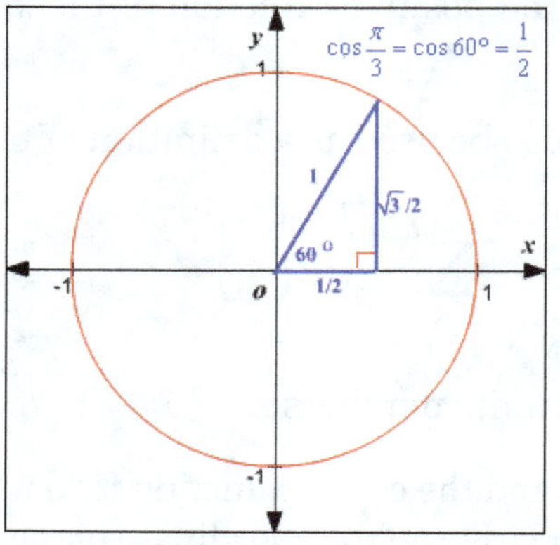

You can plot these points on a coordinate plane to show part of the cosine function, the part between 0 and 2π.

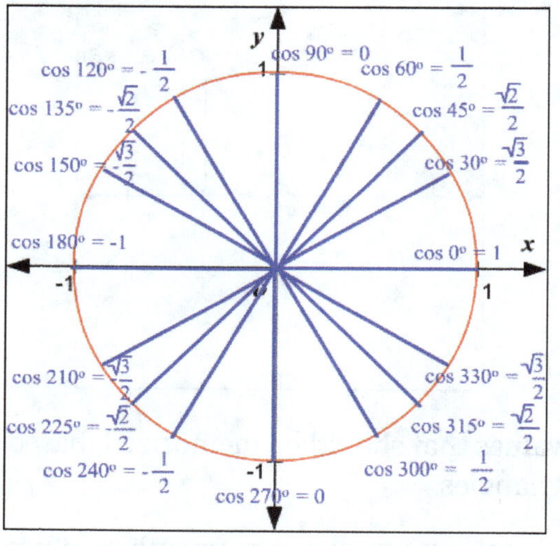

For values of θ less than 0 or greater than 2π you can find the value of $\cos(\theta)$ using the reference angle.

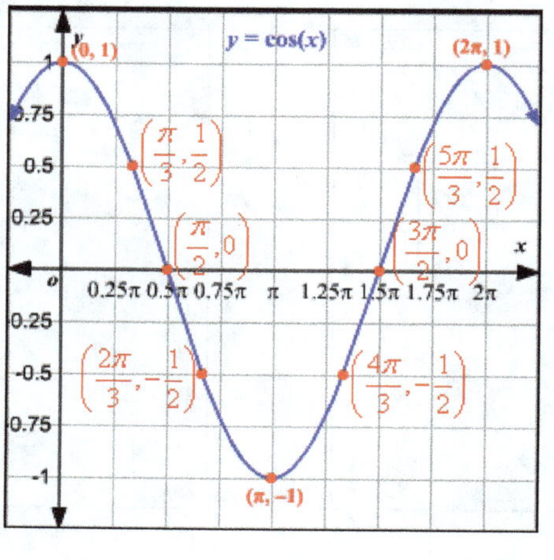

The graph of the function over a wider interval is shown below:

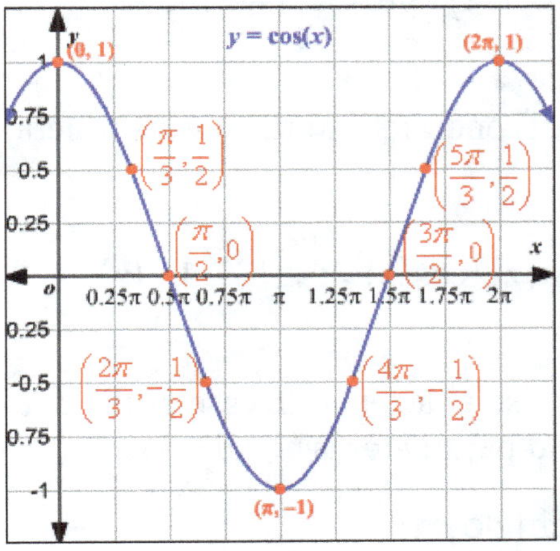

Note that the of the function is the whole real line, while the range is $-1 \le y \le 1$.

The period of $f(x) = \cos(x)$ is 2π. That is, the shape of the curve repeats every 2π-unit interval on the x-axis.

The amplitude of $f(x) = \cos(x)$ is 1, that is, the height of the wave.

The modified function $y = a\cos(bx)$ has amplitude a and period $2\pi/b$.

Representation through more General Functions

The function $\cos(z)$ is a particular case of more complicated mathematical functions. For example, it is a special case of the generalized hypergeometric function ${}_0F_1(;a;w)$ with the parameter $a = \frac{1}{2}$ at $w = -\frac{z^2}{4}$:

$$\cos(z) = {}_0F_1\left(;\frac{1}{2};-\frac{z^2}{4}\right).$$

It is also a particular case of the Bessel function $J_v(z)$ with the parameter $v = \frac{1}{2}$, multiplied by $\sqrt{\frac{\pi z}{2}}$:

$$\cos(z) = \sqrt{\frac{\pi z}{2}} J_{-\frac{1}{2}}(z).$$

Other Bessel functions can also be expressed through cosine functions for similar values of the parameter:

$$\cos(z) = \sqrt{\frac{\pi i z}{2}} I_{-\frac{1}{2}}(iz) \quad \cos(z) = -\sqrt{\frac{\pi z}{2}} Y_{\frac{1}{2}}(z)$$

$$\cos(z) = \frac{1}{\sqrt{2\pi}} \left(\sqrt{iz}\, K_{\frac{1}{2}}(iz) + \sqrt{-iz}\, K_{\frac{1}{2}}(-iz) \right).$$

Struve functions can also degenerate into the cosine function for similar values of the parameter:

$$\cos(z) = 1 - \sqrt{\frac{\pi z}{2}}\, H_{\frac{1}{2}}(z) \quad \cos(z) = 1 + \sqrt{\frac{\pi i z}{2}}\, L_{\frac{1}{2}}(z).$$

But the function $\cos(z)$ is also a degenerate case of the doubly periodic Jacobi elliptic functions when their second parameter is equal to 0 or 1:

$$\cos(z) = cd(z|0) = cn(z|0)$$

$$\cos(z) = nc(iz|1) = nd(iz|1).$$

Finally, the function $\cos(z)$ is the particular case of another class of functions—the Mathieu functions:

$$\cos(z) = Ce(1, 0, z).$$

Definition of the Cosine Function for a Complex Argument

In the complex -plane, the function $\cos(z)$ is defined using the exponential function e^w in the points $w = iz$ and $w = -iz$ through the formula:

$$\cos(z) = \frac{e^{iz} + e^{-iz}}{2}.$$

The key role in this definition of $\cos(z)$ belongs to the famous Euler formula that connects the exponential, the sine, and the cosine functions:

$$e^{iz} = \cos(z) + i\sin(z).$$

Changing z to $-z$, the Euler formula can be converted into the following modification:

$$e^{iz} = \cos(z) - i\sin(z).$$

Adding the preceding formulas gives the following result:

$$\cos(z) = \frac{e^{iz} + e^{-iz}}{2}.$$

Here are two graphics showing the real and imaginary parts of the cosine function over the complex plane:

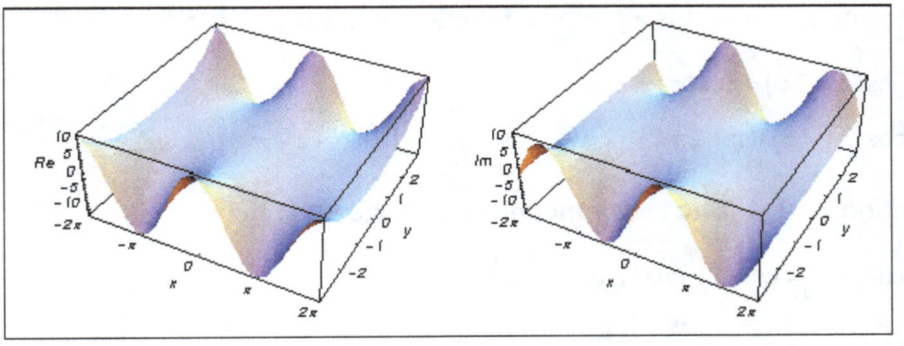

The Best-known Properties and Formulas for the Cosine Function

Values in Points

Students usually learn the following basic table of cosine function values for special points of the circle:

$$\cos(0)=1 \quad \cos\left(\frac{\pi}{6}\right)=\frac{\sqrt{3}}{2} \quad \cos=\left(\frac{\pi}{4}\right)=\frac{1}{\sqrt{2}} \quad \cos\left(\frac{\pi}{3}\right)=\frac{1}{2}$$

$$\cos\left(\frac{\pi}{2}\right)=0 \quad \cos\left(\frac{2\pi}{3}\right)=-\frac{1}{2} \quad \cos\left(\frac{3\pi}{4}\right)=-\frac{1}{\sqrt{2}} \quad \cos\left(\frac{5\pi}{6}\right)=-\frac{\sqrt{3}}{2}$$

$$\cos(\pi)=-1 \quad \cos\left(\frac{7\pi}{6}\right)=-\frac{\sqrt{3}}{2} \quad \cos\left(\frac{5\pi}{4}\right)=-\frac{1}{\sqrt{2}} \quad \cos\left(\frac{4\pi}{3}\right)=-\frac{1}{2}$$

$$\cos\left(\frac{3\pi}{2}\right)=0 \quad \cos\left(\frac{5\pi}{3}\right)=\frac{1}{2} \quad \cos\left(\frac{7\pi}{4}\right)=\frac{1}{\sqrt{2}} \quad \cos\left(\frac{11\pi}{6}\right)=\frac{\sqrt{3}}{2}$$

$$\cos(2\pi)=1 \quad \cos(\pi m)=(-1)^m /; m\in\mathbb{Z} \quad \cos\left(\pi\left(\frac{1}{2}+m\right)\right)=0/; m\in\mathbb{Z}.$$

For real values of argument z, the values of $\cos(z)$ are real.

In the points $z=2\pi n/m/; n\in\mathbb{Z}\wedge m\in\mathbb{Z}$, the values of $\cos(z)$ are algebraic. In several cases they can even be rational numbers, 0, or 1. Here are some examples:

$$\cos(0)=1 \quad \cos\left(\frac{\pi}{3}\right)=\frac{1}{2} \quad \cos\left(\frac{\pi}{2}\right)=0.$$

The values of $\cos\left(\frac{n\pi}{m}\right)$ can be expressed using only square roots if $n\in\mathbb{Z}$ and is a product of a power of 2 and distinct Fermat primes {3, 5, 17, 257, etc}.

The function $\cos(z)$ is an entire analytical function of z that is defined over the whole complex -plane and does not have branch cuts and branch points. It has an essential singular point at $z = \tilde{\infty}$. It is a periodic function with the real period 2π:

$$\cos(z + 2\pi) = \cos(z)$$

$$\cos(z) = \cos(z + 2\pi k) /; k \in \mathbb{Z} \quad \cos(z) = (-1)^k \cos(z + \pi k) /; k \in \mathbb{Z}.$$

The function $\cos(z)$ is an even function with mirror symmetry:

$$\cos(-z) = \cos(z) \quad \cos(\bar{z}) = \overline{\cos(z)}.$$

Differentiation

The derivatives of $\cos(z)$ have simple representations using either the $\sin(z)$ function or the $\cos(z)$ function:

$$\frac{\partial \cos(z)}{\partial z} = -\sin(z) \quad \frac{\partial^n \cos(z)}{\partial z^n} = \cos\left(z + \frac{\pi n}{2}\right) /; n \in \mathbb{N}^+.$$

Ordinary Differential Equation

The function $\cos(z)$ satisfies the simplest possible linear differential equation with constant coefficients:

$$w''(z) + w(z) = 0 /; w(z) = \cos(z) \wedge w(0) = 1 \wedge w'(0) = 0.$$

The complete solution of this equation can be represented as a linear combination of $\sin(z)$ and $\cos(z)$ with arbitrary constant coefficients c_1 and c_2:

$$w''(z) + w(z) = 0 /; w(z) = c_1 \cos(z) + c_2 \sin(z).$$

The function $\cos(z)$ also satisfies first-order nonlinear differential equations:

$$w'(z) - \sqrt{1 - (w(z))^2} = 0 /; w(z) = \cos(z) \wedge w(0) = 1 \wedge \operatorname{Re}(z) < \frac{\pi}{2}.$$

Series Representation

The function $\cos(z)$ has a simple series expansion at the origin that converges in the whole complex z-plane:

$$\cos(z) = 1 - \frac{z^2}{z} + \frac{z^4}{24} - \ldots = \sum_{k=0}^{\infty} \frac{(-1)^k z^{2k}}{(2k)!}.$$

For real $z = x$ this series can be interpreted as the real part of the series expansion for the exponential function e^{ix}:

$$\cos(z) = 1 - \frac{z^2}{z} + \frac{z^4}{24} - \ldots = \operatorname{Re}\left(1 + ix + \frac{(ix)^2}{2!} + \frac{(ix)^3}{3!} + \ldots\right) = \operatorname{Re}\left(\sum_{k=0}^{\infty} \frac{(ix)^k}{k!}\right) = \operatorname{Re}\left(e^{ix}\right).$$

Product Representation

The following famous infinite product representation for $\cos(z)$ clearly illustrates that $\cos(z) = 0$ at $z = \pi k - \frac{\pi}{2} \wedge k \in \mathbb{Z}$:

$$\cos(z) = \prod_{k=1}^{\infty} \left(1 - \frac{4z^2}{\pi^2 (2k-1)^2}\right).$$

Indefinite Integration

Indefinite integrals of expressions involving the cosine function can sometimes be expressed using elementary functions. However, special functions are frequently needed to express the results even when the integrands have a simple form (if they can be evaluated in closed form). Here are some examples:

$$\int \cos(z)\,dz = \sin(z)$$

$$\int \sqrt{\cos(z)}\,dz = 2E\left(\frac{z}{2} \middle| 2\right)$$

$$\int z^{\alpha-1} \cos^{v}(az)\,dz = \frac{2^{-v}(1 - v \bmod 2)}{\alpha} z^{\alpha} \binom{v}{\frac{v}{2}} -$$

$$2^{-v} z^{\alpha} \sum_{j=0}^{\left\lceil \frac{v-1}{2} \right\rceil} \binom{v}{j} \left(\Gamma(\alpha, ia(2j-v)z)(ia(2j-v)z)^{-\alpha} + \Gamma(\alpha, ia(v-2j)z)(ia(v-2j)z)^{-\alpha}\right)$$

$$/; \alpha \neq 0 \wedge v \in \mathbb{N}^{+}$$

The last integral cannot be evaluated in closed form using the known classical special functions for arbitrary values of parameters α and v.

Definite Integration

Definite integrals that contain the cosine function are sometimes simple. For example, the famous Dirichlet type and Fresnel integrals have the following values:

$$\int_0^{\infty} \frac{\cos(t) - 1}{t}\,dt = -\gamma$$

$$\int_0^{\infty}\cos\left(t^2\right)dt=\frac{1}{2}\sqrt{\frac{\pi}{2}},$$

where γ is the Euler-Mascheroni constant $\gamma = 0.577216.....$

Some special functions can be used to evaluate more complicated definite integrals. For example, elliptic integrals and gamma functions are needed to express the following integrals:

$$\int_0^{\pi}\cos^{\frac{1}{2}}(z)dz=2E(2)$$

$$\int_0^{\pi}z^{\alpha-1}\cos(z)dz=\cos\left(\frac{\pi\alpha}{2}\right)\Gamma(\alpha)/;0<\text{Re}(\alpha)<1.$$

Finite Summation

The following finite sums from the cosine can be expressed using the trigonometric functions:

$$\sum_{k=0}^{n}\cos(ak)=\csc\left(\frac{a}{2}\right)\sin\left(\frac{a(n+1)}{2}\right)\cos\left(\frac{an}{2}\right)$$

$$\sum_{k=0}^{n}(-1)^k\cos(ak)=\cos\left(\frac{1}{2}(a+n(a+\pi))\right)\sec\left(\frac{a}{2}\right)\cos\left(\frac{1}{2}n(a+\pi)\right)$$

$$\sum_{k=0}^{n}\cos(ak+z)=\csc\left(\frac{a}{2}\right)\sin\left(\frac{1}{2}a(n+1)\right)\cos\left(\frac{an}{2}+z\right)$$

$$\sum_{k=0}^{n}(-1)^k\cos(ak+z)=\cos\left(\frac{1}{2}(a+n(a+\pi))\right)\sec\left(\frac{a}{2}\right)\cos\left(\frac{1}{2}n(a+\pi)+z\right)$$

$$\sum_{k=1}^{n}z^k\cos(ka)=\frac{z\left(\cos(an)z^{n+1}-\cos(na+a)z^n-z+\cos(a)\right)}{z^2-2\cos(a)z+1}.$$

Infinite Summation

The following infinite sums can be expressed using elementary functions:

$$\sum_{k=1}^{\infty}\frac{\cos(ak)}{k}=\frac{1}{2}\log\left(\frac{1}{2(1-\cos(a))}\right)/;0<\text{Re}(a)<2\pi$$

$$\sum_{k=1}^{\infty}\frac{(-1)^{k-1}\cos(ak)}{k}=\log\left(2\cos\left(\frac{a}{2}\right)\right)/;[\text{Re}(a)]<\pi$$

$$\sum_{k=1}^{\infty}\frac{\cos(ak)}{k!}=e^{\cos(a)}\cos(\sin(a))$$

$$\sum_{k=1}^{\infty}\frac{z^k\cos(ak)}{k!}=e^{z\cos(a)}\cos(z\sin(a))$$

$$\sum_{k=1}^{\infty}z^k\cos(ak)=\frac{1-z\cos(a)}{z^2-2\cos(a)z+1}/;|z|<1.$$

Finite Products

The following finite products from the cosine can be expressed using trigonometric functions:

$$\prod_{k=1}^{2n-1}\cos\left(\frac{k\pi}{n}\right)=2^{1-2n}\left((-1)^n-1\right)/;n\in\mathbb{N}^+$$

$$\prod_{k=1}^{\left\lfloor\frac{n-1}{2}\right\rfloor}\cos\left(\frac{\pi k}{n}\right)=2^{\frac{1-n}{2}}\left(\frac{1}{2}\left(-(-1)^n\right)+\frac{1}{2}\left(1+(-1)^n\right)\sqrt{n}\right)/;n\in\mathbb{N}^+$$

$$\prod_{k=1}^{n-1}\cos\left(\frac{\pi k}{n}+z\right)=\frac{2^{1-n}}{\cos(z)}\sin\left(n\left(z+\frac{\pi}{2}\right)\right)/;n\in\mathbb{N}^+$$

$$\prod_{k=1}^{n-1}\cos\left(\frac{2\pi k}{n}+z\right)=(-2)^{1-2n}\sec(z)\left(\cos(nz)-\cos\left(\frac{n\pi}{2}\right)\right)/;n\in\mathbb{N}^+.$$

Infinite Products

The following infinite product that contains the cosine function can be expressed using the sine function:

$$\prod_{k=1}^{\infty}\cos\left(\frac{z}{2^k}\right)=\frac{\sin(z)}{z}.$$

Addition Formulas

The cosine of a sum can be represented by the rule: “the cosine of a sum is equal to the product of the cosines minus the product of the sines.” A similar rule is valid for the cosine of the difference:

$$\cos(a+b)=\cos(a)\cos(b)-\sin(a)\sin(b)$$

$$\cos(a-b)=\cos(a)\cos(b)+\sin(a)\sin(b).$$

Multiple Arguments

In the case of multiple arguments z, 2 z, 3 z, etc., the function $\cos(z)$ can be represented as the finite sum of terms that include powers of the sine and the cosine:

$$\cos(2z)=\cos^2(z)-\sin^2(z)$$

$$\cos(3z) = 4\cos^3(z) - 3\cos(z)$$

$$\cos(nz) = \sum_{k=0}^{\left\lfloor \frac{n}{2} \right\rfloor} (-1)^k \binom{n}{2k} \sin^{2k}(z)\cos^{n-2k}(z) /; n \in \mathbb{N}^+.$$

The function $\cos(nz)$ can also be represented as the finite sum including only the cosine of z:

$$\cos(nz) = n \sum_{k=0}^{\left\lfloor \frac{n}{2} \right\rfloor} \frac{(-1)^k (n-k-1) 2^{n-2k-1}}{k!(n-2k)!} \cos^{n-2k}(z) /; n \in \mathbb{N}^+.$$

Half Angle Formulas

The cosine of the half-angle can be represented by the following simple formula that is valid in some vertical strips:

$$\cos\left(\frac{z}{2}\right) = \sqrt{\frac{1+\cos(z)}{2}} /; |\operatorname{Re}(z)| < \pi \vee \operatorname{Re}(z)$$
$$= -\pi \wedge \operatorname{Im}(z) \geq 0 \vee \operatorname{Re}(z) = \pi \wedge \operatorname{Im}(z) \leq 0.$$

To make this formula correct for all complex z, a complicated prefactor is needed:

$$\cos\left(\frac{z}{2}\right) = c(z)\sqrt{\frac{1+\cos(z)}{2}} /; c(z) = (-1)^{\left\lfloor \frac{\operatorname{Re}(z)+\pi}{2\pi} \right\rfloor} \left(1 - \left(1 + (-1)^{\left\lfloor \frac{\operatorname{Re}(z)+\pi}{2\pi} \right\rfloor + \left\lfloor -\frac{\operatorname{Re}(z)+\pi}{2\pi} \right\rfloor}\right) \theta(-\operatorname{Im}(z))\right),$$

where $c(z)$ contains the unit step, real part, imaginary part, and the floor functions.

Sums of Two Direct Functions

The sum of two cosine functions can be described by the rule: "the sum of the cosines is equal to two times the cosine of the half-difference multiplied by the cosine of the half-sum."

A similar rule is valid for the difference of two cosines:

$$\cos(a) + \cos(b) = 2\cos\left(\frac{a-b}{2}\right)\cos\left(\frac{a+b}{2}\right)$$

$$\cos(a) - \cos(b) = -2\sin\left(\frac{a+b}{2}\right)\sin\left(\frac{a-b}{2}\right).$$

Products involving the Direct Function

The product of two cosine functions and the product of the cosine and sine have the following representations:

$$\cos(a)\cos(b)=\frac{1}{2}(\cos(a-b)+\cos(a+b))$$
$$\cos(a)\sin(b)=\frac{1}{2}(\sin(b-a)+\sin(a+b)).$$

Powers of the Direct Function

The integer powers of the cosine functions can be expanded as finite sums of cosine functions with multiple arguments. These sums include binomial coefficients:

$$\cos^n(z)=2^{-n}\binom{n}{\frac{n}{2}}(1-n \bmod 2)+2^{1-n}\sum_{k=0}^{\left\lfloor\frac{n-1}{2}\right\rfloor}\binom{n}{k}\cos((n-2k)z)/;n\in\mathbb{N}^+.$$

Inequalities

The best-known inequalities for cosine functions are the following:

$$|\cos(x)|\leq 1/;x\in\mathbb{R};\quad \cos(x)<\frac{\sin(x)}{x}/;0<x<4.493409.$$

Relations with its Inverse Function

There are simple relations between the function $\cos(z)$ and its inverse function $\cos^{-1}(z)$:

$$\cos^{-1}(\cos(z))=z\ \ \cos^{-1}(\cos(z))=z/;0<\mathrm{Re}(z)<\pi\vee\mathrm{Re}(z)$$
$$=0\wedge\mathrm{Im}(z)\geq 0\vee\mathrm{Re}(z)=\pi\wedge\mathrm{Im}(z)\leq 0.$$

The second formula is valid at least in the vertical strip $0<\mathrm{Re}(z)<\pi$. Outside of this strip, a much more complicated relation (that contains the unit step, real part, and the floor functions) holds:

$$\cos^{-1}(\cos(z))=\frac{\pi}{2}\left(1-(-1)^{\left\lfloor-\frac{\mathrm{Re}(z)}{\pi}\right\rfloor}\right)+(-1)^{\left\lfloor-\frac{\mathrm{Re}(z)}{\pi}\right\rfloor}\left(\left(1+(-1)^{\left\lfloor\frac{\mathrm{Re}(z)}{\pi}\right\rfloor+\left\lfloor\frac{\mathrm{Re}(z)}{\pi}\right\rfloor}\right)\theta(\mathrm{Im}(z))-1\right)$$
$$\left(z+\pi\left\lfloor-\frac{\mathrm{Re}(z)}{\pi}\right\rfloor\right).$$

Representations through other Trigonometric Functions

Cosine and sine functions are connected by a very simple formula including the linear function in the argument:

$$\cos(z) = \sin\left(\frac{\pi}{2} - z\right).$$

Another famous formula, connecting $\cos(z)$ and $\sin(z)$, is shown in the well-known Pythagorean theorem:

$$\cos^2(z) = 1 - \sin^2(z)$$

$$\cos(z) = \sqrt{1 - \sin^2(z)} \,/;\, |\operatorname{Re}(z)| < \frac{\pi}{2} \vee \operatorname{Re}(z) = -\frac{\pi}{2} \wedge \operatorname{Im}(z) \geq 0 \vee \operatorname{Re}(z) = \frac{\pi}{2} \wedge \operatorname{Im}(z) \leq 0.$$

The last restriction on z can be removed, but the formula will get a complicated coefficient $c(z)$ with $|c(z)| = 1$, that contains the unit step, real part, imaginary part, and the floor function:

$$\cos(z) = c(z)\sqrt{1 - \sin^2(z)} \,/;\, c(z) = (-1)^{\left\lfloor \frac{1}{2} - \frac{\operatorname{Re}(z)}{\pi} \right\rfloor \left(1 - \left(1 + (-1)^{\left\lfloor \frac{\operatorname{Re}(z)}{\pi} - \frac{1}{2} \right\rfloor + \left\lfloor \frac{1}{2} - \frac{\operatorname{Re}(z)}{\pi} \right\rfloor}\right) \theta(\operatorname{Im}(z))\right)}.$$

The cosine function can also be represented using other trigonometric functions by the following formulas:

$$\cos(z) = \frac{1 - \tan^2\left(\frac{z}{2}\right)}{1 + \tan^2\left(\frac{z}{2}\right)} \quad \cos(z) = \frac{\cot^2\left(\frac{z}{2}\right) - 1}{\cot^2\left(\frac{z}{2}\right) + 1}$$

$$\cos(z) = \frac{1}{\csc\left(\frac{\pi}{2} - z\right)} \quad \cos(z) = \frac{1}{\sec(z)}.$$

Representations through Hyperbolic Functions

The cosine function has representations using the hyperbolic functions:

$$\cos(z) = \cosh(iz) \quad \cos(iz) = \cosh(z) \quad \cos(z) = -i\sinh\left(\frac{\pi i}{2} - iz\right)$$

$$\cos(z) = \frac{1+\tanh^2\left(\frac{zi}{2}\right)}{1-\tanh^2\left(\frac{zi}{2}\right)} \quad \cos(z) = \frac{\coth^2\left(\frac{iz}{2}\right)+1}{\coth^2\left(\frac{iz}{2}\right)-1} \quad \cos(z)$$

$$= -\frac{i}{\operatorname{csch}\left(\frac{\pi i}{2}-iz\right)} \quad \cos(z) = \frac{1}{\operatorname{sech}(iz)}$$

Applications

The cosine function is used throughout mathematics, the exact sciences, and engineering.

TANGENT

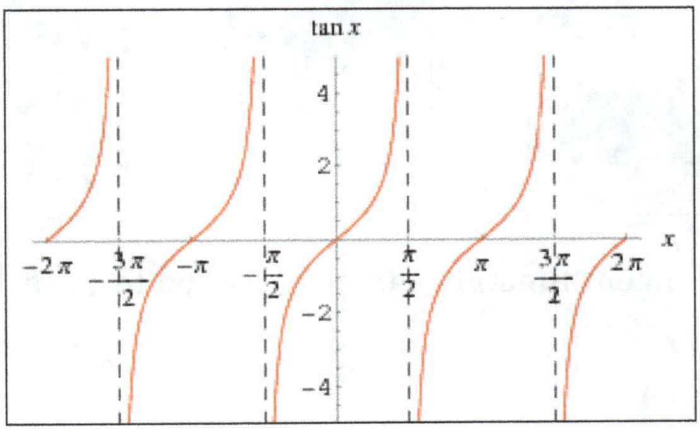

The tangent function is defined by,

$$\tan x = \frac{\sin x}{\cos x},$$

where sin x is the sine function and cos x is the cosine function. The notation tg x is sometimes also used.

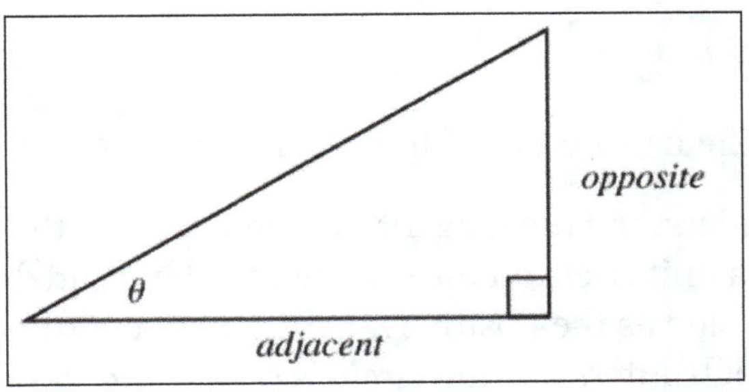

The common schoolbook definition of the tangent of an angle theta in a right triangle (which is equivalent to the definition just given) is as the ratio of the side lengths opposite to the angle and adjacent the angle, i.e.,

$$\tan\theta = \frac{\text{opposite}}{\text{adjacent}}.$$

The word "tangent" also has an important related meaning as a line or plane which touches a given curve or solid at a single point. These geometrical objects are then called a tangent line or tangent plane, respectively.

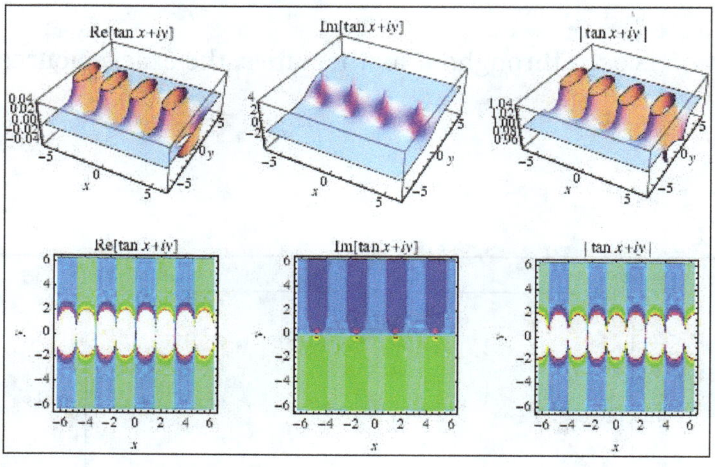

The definition of the tangent function can be extended to complex arguments z using the definition,

$$\begin{aligned}\tan z &= \frac{i\left(e^{-iz}-e^{iz}\right)}{e^{-iz}+e^{iz}}\\ &= \frac{e^{iz}-e^{-iz}}{i\left(e^{iz}+e^{-iz}\right)}\\ &= \frac{i\left(1-e^{2iz}\right)}{1+e^{2iz}}\\ &= \frac{e^{2iz}-1}{i\left(e^{2iz}+1\right)},\end{aligned}$$

where e is the base of the natural logarithm and i is the imaginary number.

The simplest way to understand the tangent function is to use the unit circle. For a given angle measure θ draw a unit circle on the coordinate plane and draw the angle centered at the origin, with one side as the positive x-axis. The x -coordinate of the point where the other side of the angle intersects the circle is cos(θ) and the y -coordinate is sin(θ).

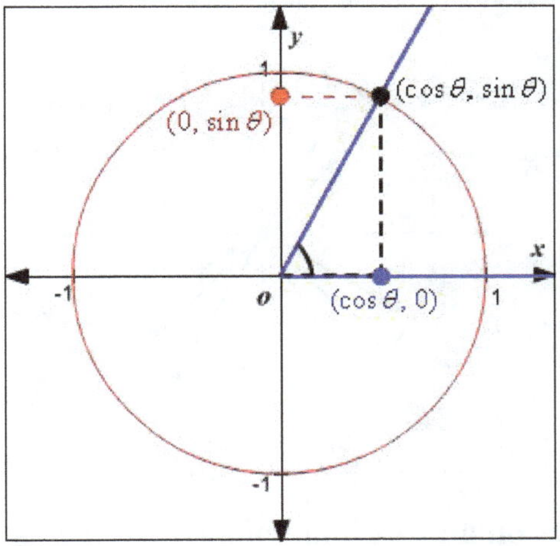

There are a few sine and cosine values that should be memorized, based on 30°–60°–90° triangles and 45°–45°–90° triangles. Based on these, you can work out the related values for tangent.

$\sin(\theta)$	$\cos(\theta)$	$\tan(\theta)$
$\sin(0°)=0$	$\cos(0°)=1$	$\tan(0°)=\frac{0}{1}=0$
$\sin(30°)=\frac{1}{2}$	$\cos(30°)=\frac{\sqrt{3}}{2}$	$\tan(30°)=\frac{1}{2}\cdot\frac{2}{\sqrt{3}}=\frac{\sqrt{3}}{2}$
$\sin(45°)=\frac{\sqrt{2}}{2}$	$\cos(45°)=\frac{\sqrt{2}}{2}$	$\tan(45°)=\frac{\sqrt{2}}{2}\cdot\frac{2}{\sqrt{2}}=1$
$\sin(60°)=\frac{\sqrt{3}}{2}$	$\cos(60°)=\frac{1}{2}$	$\tan(60°)=\frac{\sqrt{3}}{2}\cdot\frac{2}{1}=\sqrt{3}$
$\sin(90°)=1$	$\cos(90°)=0$	$\tan(90°)=\frac{1}{0}=\text{undef.}$

- For angles with their terminal arm in Quadrant II, since sine is positive and cosine is negative, tangent is negative.
- For angles with their terminal arm in Quadrant III, since sine is negative and cosine is negative, tangent is positive.
- For angles with their terminal arm in Quadrant IV, since sine is negative and cosine is positive, tangent is negative.

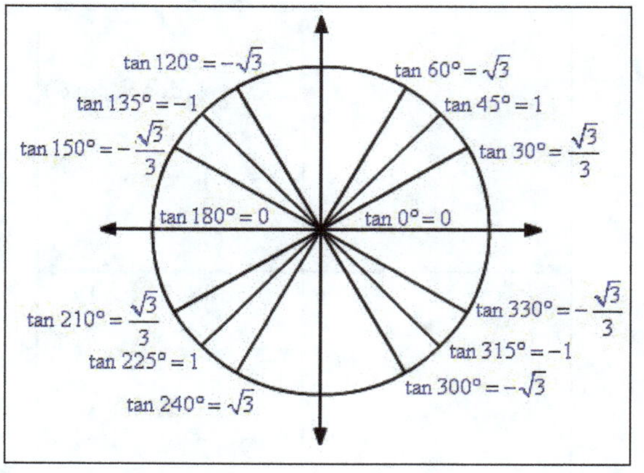

You can plot these points on a coordinate plane to show part of the function, the part between 0 and 2π.

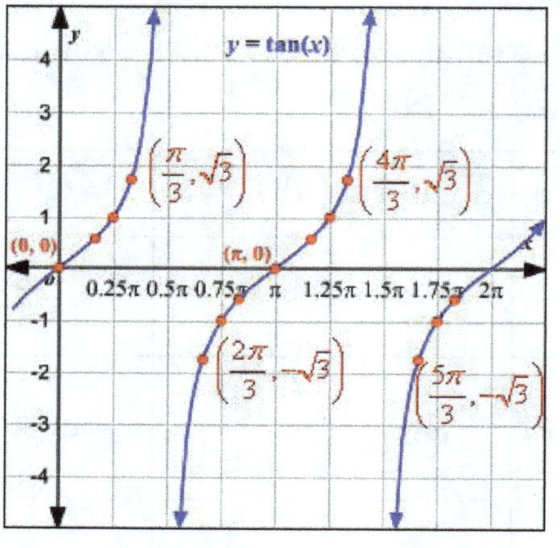

For values of θ less than 0 or greater than 2π you can find the value of θ using the reference angle.

The graph of the function over a wider interval is shown below:

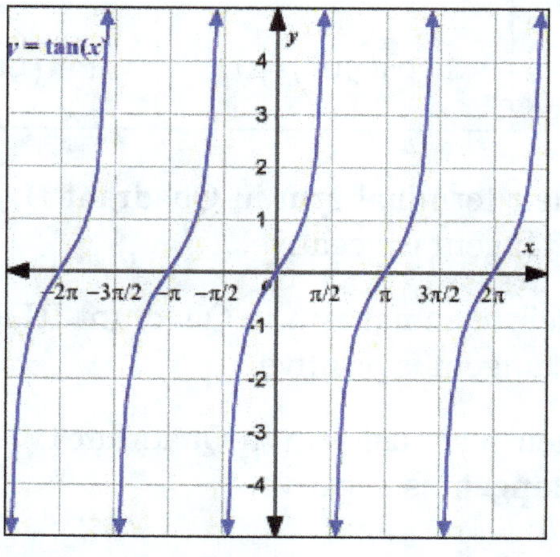

Note that the domain of the function is the whole real line and the range is $-\infty \leq y \leq \infty$.

A related function known as the hyperbolic tangent is similarly defined,

$$\tanh z = \frac{e^z - e^{-z}}{e^z + e^{-z}}.$$

An important tangent identity is given by,

$$\tan^2 \theta + 1 = \sec^2 \theta.$$

Angle addition, subtraction, half-angle, and multiple-angle formulas are given by,

$$\tan(\alpha+\beta) = \frac{\tan\alpha + \tan\beta}{1 - \tan\alpha\tan\beta}$$

$$\tan(\alpha-\beta) = \frac{\tan\alpha - \tan\beta}{1 + \tan\alpha\tan\beta}$$

$$\tan(2\alpha) = \frac{2\tan\alpha}{1 - \tan^2\alpha}$$

$$\tan(n\alpha) = \frac{\tan\left[(n-1)\alpha\right] + \tan\alpha}{1 - \tan\left[(n-1)\alpha\right]\tan\alpha}$$

$$\tan\left(\frac{\alpha}{2}\right) = \frac{\sin\alpha}{1+\cos\alpha}$$

$$= \frac{1-\cos\alpha}{\sin\alpha}$$

$$= \frac{\tan\alpha\sin\alpha}{\tan\alpha + \sin\alpha}.$$

The sine and cosine functions can conveniently be expressed in terms of a tangent as,

$$\cos t = \frac{1 - \tan^2\left(\frac{1}{2}t\right)}{1 + \tan^2\left(\frac{1}{2}t\right)}$$

$$\sin t = \frac{2\tan\left(\frac{1}{2}t\right)}{1 + \tan^2\left(\frac{1}{2}t\right)},$$

which can be particularly convenient in polynomial computations such as Gröbner basis since it reduces the number of equations compared with explicit inclusion of $\cos t$ and $\sin t$ together with the additional relation $\cos^2 t+\sin^2 t-1=0$.

These lead to the pretty identity,

$$\tan\left(x+\frac{1}{4}\pi\right)=\frac{1+\tan x}{1-\tan x}.$$

There is also a beautiful angle addition identity for three variables,

$$\tan(\alpha+\beta+\gamma)=\frac{\tan\alpha+\tan\beta+\tan\gamma-\tan\alpha\tan\beta\tan\gamma}{1-\tan\beta\tan\gamma-\tan\gamma\tan\alpha-\tan\alpha\tan\beta}.$$

Another tangent identity is,

$$\begin{aligned}\tan(nx)&=\frac{\sum_{k=0}^{n}(-1)^k\binom{n}{2k+1}t^{2k+1}}{\sum_{k=0}^{n}(-1)^k\binom{n}{2k}t^{2k}}\\&=\frac{\frac{1}{2}i(1-it)^n-(1+it)^n}{\frac{1}{2}i(1-it)^n-(1+it)^n}\\&=\frac{1}{i}\frac{(1+it)^n-(1-it)^n}{(1+it)^n+(1-it)^n},\end{aligned}$$

where t = tan x. Written explicitly,

$$\tan(nx)=\frac{2i(1-i\tan t)^n}{(1-i\tan t)^n+(1+i\tan t)^n}-i,$$

This gives the first few expansions as,

$$\begin{aligned}\tan x&=t\\\tan(2x)&=\frac{2t}{1-t^2}\\\tan(3x)&=\frac{3t-t^3}{1-3t^2}\\\tan(4x)&=\frac{4t-4t^3}{1-6t^2+r^4}\\\tan(5x)&=\frac{5t-10t^3+t^5}{1-10t^2+5t^4}.\end{aligned}$$

A beautiful formula that generalizes the tangent angle addition formula, $\tan(4x)=\frac{4t-4t^3}{1-6t^2+r^4}$, and $\tan(5x)=\frac{5t-10t^3+t^5}{1-10t^2+5t^4}$ is given by,

$$\tan\left(\sum_{n=1}^{N}\theta_n\right)=i\frac{\prod_{n=1}^{N}(1-i\tan\theta_n)-\prod_{n=1}^{N}(i\tan\theta_n+1)}{\prod_{n=1}^{N}(1-i\tan\theta_n+1)-\prod_{n=1}^{N}(1-i\tan\theta_n)}$$

There are a number of simple but interesting tangent identities based on those given above, including,

$$\tan(A+60°)\tan(A-60°)+\tan A\tan(A+60°)+\tan A\tan(A-60°)=-3$$

The Maclaurin series valid for $-\pi/2<x<\pi/2$ for the tangent function is,

$$\tan x=\sum_{n=1}^{\infty}\frac{(-1)^{n-1}2^{2n}\left(2^{2n}-1\right)B_{2n}}{(2n)!}x^{2n-1}$$
$$=x+\frac{1}{3}x^3+\frac{2}{15}x^5+\frac{17}{315}x^7+\frac{62}{2835}x^9+\ldots$$

where B_n is a Bernoulli number.

tan x is irrational for any rational $x\neq 0$, which can be proved by writing tan x as a continued fraction as,

$$\tan x=\cfrac{x}{1-\cfrac{x^2}{3-\cfrac{x^2}{5-\cfrac{x^2}{7-\ldots}}}}$$

and

$$\tan x=\cfrac{1}{\frac{1}{x}-\cfrac{1}{\frac{3}{x}-\cfrac{1}{\frac{5}{x}-\cfrac{1}{\frac{7}{x}-\ldots}}}}$$

both due to Lambert.

An interesting identity involving the product of tangents is,

$$\prod_{k=1}^{[(n-1)/2]} \tan\left(\frac{k\pi}{n}\right) = \begin{cases} \sqrt{n} & \text{for n odd} \\ 1 & \text{for n even,} \end{cases}$$

where $\lfloor x \rfloor$ is the floor function.

The equation,

$$x = \tan x,$$

which is equivalent to $\tan(x) = 1$, where $\text{tanc}(x)$ is the tanc function, does not have simple closed-form solutions.

The difference between consecutive solutions gets closer and closer to π for higher order solutions. The function $\text{tanc}\, x \equiv (\tan x)/x$ is sometimes known as the tanc function.

COSECANT

The cosecant function, denoted (or sometimes) is defined as the composite of the reciprocal function and the sine function. Explicitly, it is defined as follows:

$$\csc x := \frac{1}{\sin x}, \qquad \sin x \neq 0$$

The cosecant function is not defined at integer multiples of π, because the sine function takes the value zero at these points.

Key Data

Item	Value
Default domain	All reals except integer multiples of π.
Range	$(-\infty,-1]\cup[1,\infty)$, i.e., $\{y \mid \lvert y\rvert \geq 1\}$ no absolute minimum value or absolute maximum value.
Period	2π
Vertical asymptotes	All lines of the form $x = n\pi, n$ varying over integers. For n even, the left hand limit is $-\infty$ and the right hand limit is $+\infty$. For n odd, the left hand limit is $+\infty$ and the right hand limit is $-\infty$.
Local minimum values and points of attainment	Local minimum value of 1, attained at $x = 2n\pi + \pi/2, n$ varying over integers.
Local maximum values and points of attainment	Local maximum value of -1, attained at, $x = 2n\pi - \pi/2, n$ varying over integers.

Points of inflection	None.
First derivative	$x \mapsto -\csc x \cot x$
Antiderivative	$x \mapsto \ln\lvert \csc x - \cot x \rvert + C = -\ln\lvert \csc x + \cot x \rvert + C$

Like the other trigonometric functions, the cosecant can be represented as a line segment associated with the unit circle. The diagram shows the cosecant for an angle of rotation θ of forty-five degrees (measured anti-clockwise from the positive x-axis). Line segment AF lies on the line that is tangent to the circle at point A. Line segment OF (shown in red) represents the cosecant, and is an extension of line segment OP. Remember that point P is a point on the circumference of the circle whose x and y coordinates represent the value of cos (θ) and sin (θ) respectively (the line segments representing the sine and cosine are also shown). As the value of sin (θ) approaches one, so does the value of csc (θ). As the value of sin (θ) approaches zero, however, the value of csc (θ) tends to infinity. The value of csc (θ) when sin (θ) equals zero is thus said to be undefined.

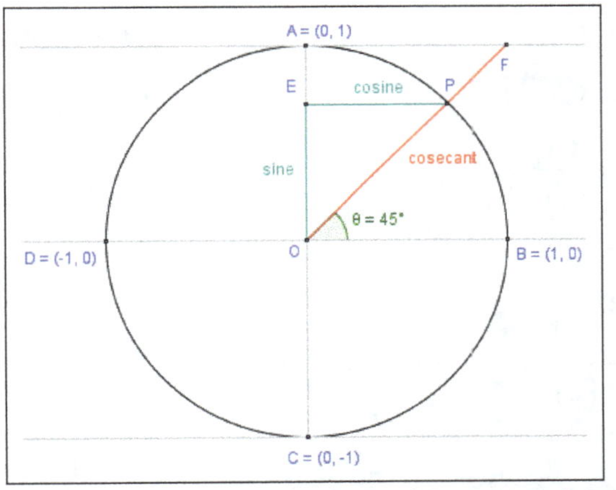

The cosecant as a line segment of the unit circle.

Here is a graphic of the cosecant function $f(x) = \csc(x)$ for real values of its argument x.

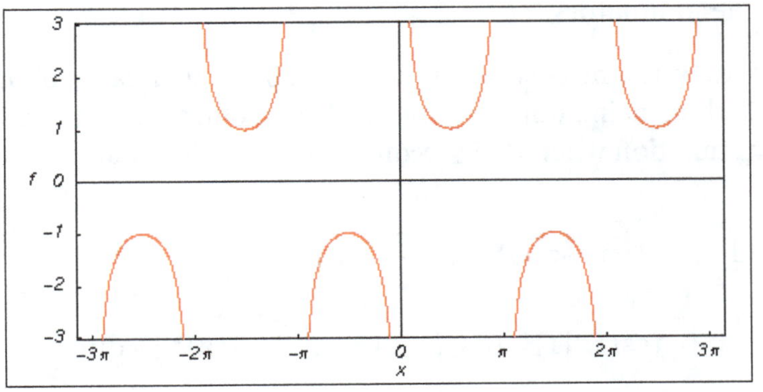

Representation through more General Functions

The cosecant function $\csc(z)$ can be represented using more general mathematical functions. As the ratio of one divided by the sine function that is a particular case of the generalized hypergeometric, Bessel, Struve, and Mathieu functions, the cosecant function can also be represented as ratios of one and those special functions. Here are some examples:

$$\csc(z) = \frac{1}{z\,{}_0F_1\left(;\frac{3}{2};-\frac{z^2}{4}\right)}$$

$$\csc(z) = \sqrt{\frac{2}{\pi}}\frac{1}{\sqrt{z}J_{\frac{1}{2}}(z)}$$

$$\csc(z) = \sqrt{\frac{2}{\pi}}\frac{i}{\sqrt{iz}I_{\frac{1}{2}}(iz)}$$

$$\csc(z) = \sqrt{\frac{2}{\pi}}\frac{1}{\sqrt{z}Y_{-\frac{1}{2}}(z)}$$

$$\csc(z) = \frac{\sqrt{\frac{2}{\pi}}}{\sqrt{z}H_{-\frac{1}{2}}(z)}$$

$$\csc(z) = \frac{1}{Se(1,0,z)}.$$

But these representations are not very useful because they include complicated special functions in the denominators.

It is more useful to write the cosecant function as particular cases of one special function. That can be done using doubly periodic Jacobi elliptic functions that degenerate into the cosecant function when their second parameter is equal to 0 or 1:

$$\csc(z) = \mathrm{ds}(z\,|\,0) = \mathrm{ns}(z\,|\,0) = \mathrm{dc}\left(\frac{\pi}{2}-z\,|\,0\right) =$$

$$\mathrm{nc}\left(\frac{\pi}{2}-z\,|\,0\right) = i\,\mathrm{cs}(iz\,|\,1) = i\,\mathrm{ds}(iz\,|\,1) = \mathrm{cn}\left(\frac{\pi i}{2}-iz\,|\,1\right) = \mathrm{dn}\left(\frac{\pi i}{2}-iz\,|\,1\right)$$

Definition of the Cosecant Function for a Complex Argument

In the complex z -plane, the function csn(z) is defined using sin(z) or the exponential function e^w in the points iz and -iz through the formula:

$$\csc(z) = \frac{1}{\sin(z)} = \frac{2i}{e^{iz} - e^{-iz}}$$

In the points $z = \pi k/; k \in \mathbb{Z}$, where sin(z) is zero, the denominator of the last formula equals zero and csc(z) has singularities (poles of the first order).

Here are two graphics showing the real and imaginary parts of the cosecant function over the complex plane.

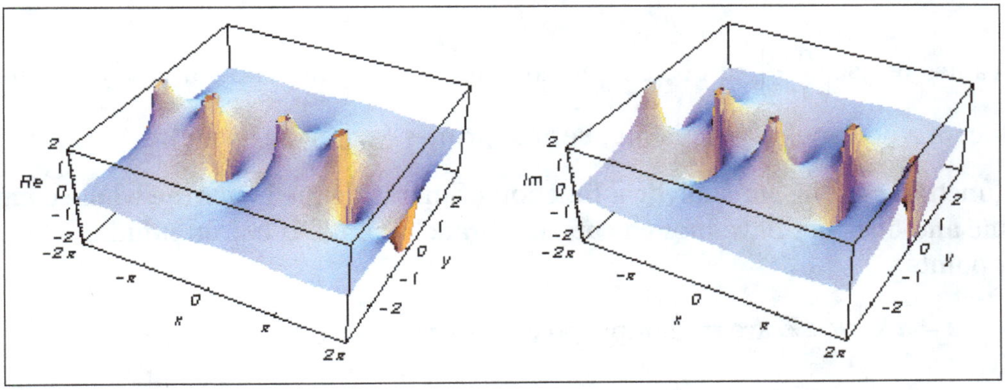

The Best-known Properties and Formulas for the Cosecant Function

Values in Points

Using the connection between the sine and cosecant functions, the following table of cosecant function values for angles between 0 and 2π can be derived:

$$\csc(0) = \overline{\infty} \quad \csc\left(\frac{\pi}{6}\right) = 2 \quad \csc\left(\frac{\pi}{4}\right) = \sqrt{2} \quad \csc\left(\frac{\pi}{3}\right) = \frac{2}{\sqrt{3}}$$

$$\csc\left(\frac{\pi}{2}\right) = 1 \quad \csc\left(\frac{2\pi}{3}\right) = \frac{2}{\sqrt{3}} \quad \csc\left(\frac{3\pi}{4}\right) = \sqrt{2} \quad \csc\left(\frac{5\pi}{6}\right) = 2$$

$$\csc(\pi) = \overline{\infty} \quad \csc\left(\frac{7\pi}{6}\right) = -2 \quad \csc\left(\frac{5\pi}{4}\right) = -\sqrt{2} \quad \csc\left(\frac{4\pi}{3}\right) = -\frac{2}{\sqrt{3}}$$

$$\csc\left(\frac{3\pi}{2}\right) = -1 \quad \csc\left(\frac{5\pi}{3}\right) = -\frac{2}{\sqrt{3}} \quad \csc\left(\frac{7\pi}{4}\right) = -\sqrt{2} \quad \csc\left(\frac{11\pi}{6}\right) = -2$$

$$\csc(2\pi)=\tilde{\infty} \quad \csc(\pi m)=\tilde{\infty}/;m\in\mathbb{Z} \quad \csc\left(\pi\left(\frac{1}{2}+m\right)\right)=(-1)^m /;m\in\mathbb{Z}$$

General Characteristics

For real values of argument z, the values of csc(z) are real.

In the points $z=2\pi n/m/;n\in\mathbb{Z}\wedge m\in\mathbb{Z}$ the values of csc(z) are algebraic. In several cases they can be integers -2, -1, 1 or 2:

$$\csc\left(-\frac{\pi}{2}\right)=-1 \quad \csc\left(-\frac{\pi}{6}\right)=-2 \quad \csc\left(\frac{\pi}{2}\right)=1 \quad \csc\left(\frac{\pi}{6}\right)=2$$

The values of $\csc\left(\frac{n\pi}{m}\right)$ can be expressed using only square roots if $n\in\mathbb{Z}$ and m is a product of a power of 2 and distinct Fermat primes {3, 5, 17, 257, etc}.

The function csc(z) is an analytical function of that is defined over the whole complex z-plane and does not have branch cuts and branch points. It has an infinite set of singular points:

- $z=\pi k/;k\in\mathbb{Z}$ are the simple poles with residues $(-1)^k$.
- $z=\tilde{\infty}$ is an essential singular point.

It is a periodic function with the real period 2π:

$$\csc(z+2\pi)=\csc(z)$$

$$\csc(z)=\csc(z+2\pi k)/;k\in\mathbb{Z} \quad \csc(z)=(-1)^k\csc(z+\pi k)/;k\in\mathbb{Z}$$

The function csc(z) is an odd function with mirror symmetry:

$$\csc(-z)=-\csc(z) \quad \csc(\bar{z})=\overline{\csc(z)}$$

Differentiation

The first derivative of csc(z) has simple representations using either the cot(z) function or the csc(z) function:

$$\frac{\partial\csc(z)}{\partial z}=-\cot(z)\csc(z)$$

The nth derivative of csc(z) has much more complicated representations than symbolic n[th] derivatives for sin(z) and cos(z):

$$\frac{\partial^n \csc(z)}{\partial z^n} = \csc(z)\left(\delta_n + (n+1)!\sum_{k=0}^{n}\sum_{j=0}^{\left\lceil \frac{k-1}{2} \right\rceil}\frac{(-1)^{j+n}\, 2^{1-k}(k-2j)^n \csc^k(z)}{(k+1)\, j!(k-j)!(n-k)!}\cos\left(\frac{\pi(k-n)}{2}+(k-2j)z\right)\right)/;n\in\mathbb{N}$$

where δ_n is the Kronecker delta symbol: $\delta_0 = 1$ and $\delta_n = 0/;n \neq 0$.

Ordinary Differential Equation

The function csc(z) satisfies the following first-order nonlinear differential equation:

$$w'(z)^2 - w(z)^4 + w(z)^2 = 0/;\ w(z) = \csc(z)$$

Series Representation

The function csc(z) has the following Laurent series expansion at the origin that converges for all finite values z with $0 < |z| < \pi$:

$$\csc(z) = \frac{1}{z} + \frac{z}{6} + \frac{7z^3}{360} + \ldots = \frac{1}{z} + \sum_{k=1}^{\infty}\frac{(-1)^{k-1}\, 2\left(2^{2k-1}-1\right)B_{2k}z^{2k-1}}{(2k)!}$$

where B_{2K} are the Bernoulli numbers.

The cosecant function csc(z) can also be represented using other kinds of series by the following formulas:

$$\csc(z) = \frac{1}{z} + 2z\sum_{k=1}^{\infty}\frac{(-1)^k}{z^2 - \pi^2 k^2}/;\ \frac{z}{\pi} \notin \mathbb{Z}$$

$$\csc^2(z) = \sum_{k=-\infty}^{\infty}\frac{1}{(z-\pi k)^2}/;\ \frac{z}{\pi} \notin \mathbb{Z}$$

Integral Representation

The function csc(z) has well-known integral representation through the following definite integral along the positive part of the real axis:

$$\csc(z) = \frac{1}{\pi}\int_0^{\infty}\frac{1}{t^2+t}t^{z/\pi}dt/;0 < \mathrm{Re}(z) < \pi$$

Product Representation

The famous infinite product representation for can be easily rewritten as the following product representation for the cosecant function:

$$\csc(z) = \lim_{n\to\infty} \sum_{k=-n}^{n} \frac{(-1)^k}{z - \pi k} /; \frac{z}{\pi} \in \mathbb{Z}$$

Indefinite Integration

Indefinite integrals of expressions that contain the cosecant function can sometimes be expressed using elementary functions. However, special functions are frequently needed to express the results even when the integrands have a simple form (if they can be evaluated in closed form). Here are some examples:

$$\int \csc(z)dz = \log\left(\tan\left(\frac{z}{2}\right)\right)$$

$$\int \sqrt{\csc(z)}\,dz = -2\csc^{\frac{1}{2}}(z)\sin^{\frac{1}{2}}(z)F\left(\frac{1}{4}(\pi - 2z)|2\right)$$

$$\int \csc^{v}(az)dz = -\frac{\cos(az)\csc^{v-1}(az)\sin^{2}(az)^{\frac{v-1}{2}}}{a}{}_2F_1\left(\frac{1}{2},\frac{v+1}{2};\frac{3}{2};\cos^2(az)\right)$$

Definite Integration

Definite integrals that contain the cosecant function are sometimes simple. For example, the famous Catalan constant can be defined as the value of the following integral:

$$\frac{1}{2}\int_0^{\frac{\pi}{2}} t\csc(t)dt = C$$

This constant also appears in the following integral:

$$\int_0^{\frac{\pi}{2}} t^2 \csc(t)dt = 2C\pi - \frac{7\zeta(3)}{2}$$

Some special functions can be used to evaluate more complicated definite integrals. For example, polylogarithmical, zeta, and gamma functions are needed to express the following integrals:

$$\int_0^{\frac{\pi}{3}} t^3 \csc^2(t)dt = \frac{1}{162}\begin{pmatrix} 6i\pi^3 - 2\sqrt{3}\pi^3 + 54\pi^2\log\left(-\sqrt[3]{-1}\left(-1+(-1)^{\frac{2}{3}}\right)\right) \\ +162i\pi\,Li_2\left(-\sqrt[3]{-1}\right) + 243\,Li_3\left(-\sqrt[3]{-1}\right) - 243\zeta(3) \end{pmatrix}$$

$$\int_0^{\frac{\pi}{2}} \csc^a(t)\,dt = \frac{\sqrt{\pi}}{2\Gamma\left(1-\frac{a}{2}\right)}\Gamma\left(\frac{1-a}{2}\right)/;\mathrm{Re}(a)<1$$

Finite Summation

The following finite sums that contain the cosecant function have simple values:

$$\sum_{k=0}^{n-1} \csc^2\left(\frac{\pi k}{n}+z\right) = n^2 \csc^2(nz)/;n\in\mathbb{N}^+$$

$$\sum_{k=1}^{n-1} \csc^2\left(\frac{k\pi}{n}\right) = \frac{1}{3}\left(n^2-1\right)/;n\in\mathbb{N}^+$$

$$\sum_{k=1}^{n} \csc^2\left(\frac{k\pi}{2n+1}\right) = \frac{2}{3}n(n+1)/;n\in\mathbb{N}^+$$

$$\sum_{k=1}^{\left\lfloor\frac{n-1}{2}\right\rfloor} \csc^2\left(\frac{k\pi}{2n}\right) = \frac{1}{4}\left(2n^2-3(-1)^n-5\right)/;n\in\mathbb{N}^+$$

$$\sum_{k=0}^{\left\lfloor\frac{n-2}{2}\right\rfloor} \csc^2\left(\frac{(2k+1)\pi}{2n}\right) = \frac{1}{4}\left(2n^2+(-1)^n-1\right)/;n\in\mathbb{N}^+$$

$$\sum_{k=0}^{n} \csc\left(\frac{z}{2^k}\right) = \cot\left(\frac{z}{2^{n+1}}\right)-\cot(z)/;n\in\mathbb{N}^+$$

Infinite Summation

The following infinite sum that contains the cosecant has a simple value:

$$\sum_{k=1}^{\infty} \frac{\csc\left(k\pi\sqrt{2}\right)}{k^3} = \frac{13\pi^3}{360\sqrt{2}}$$

Finite Products

The following finite product from the cosecant can also be represented using the cosecant function:

$$\prod_{k=0}^{n-1} \frac{\csc\left(k\pi\sqrt{2}\right)}{k^3} = \frac{13\pi^3}{360\sqrt{2}}$$

Addition Formulas

The cosecant of a sum and the cosecant of a difference can be represented by the formulas that follow from corresponding formulas for the sine of a sum and the sine of a difference:

$$\csc(a+b)=\frac{1}{\cos(b)\sin(a)+\cos(a)\sin(b)}$$

$$\csc(a-b)=\frac{1}{\cos(b)\sin(a)-\cos(a)\sin(b)}$$

Multiple Arguments

In the case of multiple arguments z, 2z, 3z, etc., the function csc(z) can be represented as a rational function that contains powers of cosecants and secants. Here are two examples:

$$\csc(2z)=\frac{1}{2}\csc(z)\sec(z)$$

$$\csc(3z)=\frac{\csc^3(z)}{3\csc^2(z)-4}$$

Half-angle Formulas

The cosecant of a half-angle can be represented by the following simple formula that is valid in a vertical strip:

$$\csc\left(\frac{z}{2}\right)=\frac{c(z)}{\sqrt{1-\cos(z)}}\ /;\ c(z)=\sqrt{2}(-1)^{\left[\frac{\operatorname{Re}(z)}{2\pi}\right]}\left(1-\left(1+(-1)^{\left[\frac{\operatorname{Re}(z)}{2\pi}\right]+\left[-\frac{\operatorname{Re}(z)}{2\pi}\right]}\right)\theta(-\operatorname{Im}(z))\right)$$

Where c(z) contains the unit step, real part, imaginary part, and the floor functions.

Sums of Two Direct Functions

The sum and difference of two cosecant functions can be described by the following formulas:

$$\csc(a)+\csc(b)=2\sin\left(\frac{a}{2}+\frac{b}{2}\right)\cos\left(\frac{a}{2}-\frac{b}{2}\right)\csc(a)\csc(b)$$

$$\csc(a)-\csc(b)=-2\sin\left(\frac{a}{2}-\frac{b}{2}\right)\cos\left(\frac{a}{2}+\frac{b}{2}\right)\csc(a)\csc(b)$$

Products involving the Direct Function

The product of two cosecants and the product of the cosecant and secant have the following representations:

$$\csc(a)\csc(b)=\frac{2}{\cos(a-b)-\cos(a+b)}$$

$$\csc(a)\sec(b) = \frac{2}{\sin(a-b) - \sin(a+b)}$$

Inequalities

Some inequalities for the cosecant function can be easily derived from the corresponding inequalities for the sine function:

$$\csc(x) > \frac{1}{x} /; x > 0$$

$$|\csc(x)| \geq 1 /; x \in \mathbb{R}$$

$$x\csc(x) < \frac{\pi}{2} /; -\frac{\pi}{2} < x < \frac{\pi}{2}$$

Relations with its Inverse Function

There are simple relations between the function csc(z) and its inverse function $\csc^{-1}(z)$:

$$\csc\left(\csc^{-1}(z)\right) = z \csc^{-1}\left(\csc(z)\right) = z /; |\operatorname{Re}(z)| < \frac{\pi}{2} \vee \operatorname{Re}(z) = -\frac{\pi}{2} \wedge \operatorname{Im}(z) \leq 0 \vee$$

$$\operatorname{Re}(z) = \frac{\pi}{2} \wedge \operatorname{Im}(z) \geq 0$$

The second formula is valid at least in the vertical strip $-\frac{\pi}{2} < \operatorname{Re}(z) < \frac{\pi}{2}$. Outside of this strip a much more complicated relation (that contains the unit step, real part, and the floor functions) holds:

$$\csc^{-1}\left(\csc(z)\right) = (-1)^{\left\lceil \frac{\operatorname{Re}(z)}{\pi} - \frac{1}{2} \right\rceil} \left(\left(1 + (-1)^{\left\lceil \frac{\operatorname{Re}(z)}{\pi} + \frac{1}{2} \right\rceil + \left\lceil -\frac{\operatorname{Re}(z)}{\pi} - \frac{1}{2} \right\rceil} \right) \theta(-\operatorname{Im}(z)) - 1 \right)$$

$$\left(z - \pi \left\lceil \frac{\operatorname{Re}(z)}{\pi} + \frac{1}{2} \right\rceil + \frac{\pi}{2} \left(1 + (-1)^{\left\lceil \frac{\operatorname{Re}(z)}{\pi} + \frac{1}{2} \right\rceil + \left\lceil -\frac{\operatorname{Re}(z)}{\pi} - \frac{1}{2} \right\rceil} \right) \theta(-\operatorname{Im}(z)) \right)$$

Representations through other Trigonometric Functions

Cosecant and secant functions are connected by a very simple formula that contains the linear function in the argument:

$$\csc(z) = \sec\left(\frac{\pi}{2} - z\right)$$

The cosecant function can also be represented using other trigonometric functions by the following formulas:

$$\csc(z)=\frac{1}{\cos\left(\frac{\pi}{2}-z\right)} \quad \csc(z)=\frac{\tan^2\left(\frac{z}{2}\right)+1}{2\tan\left(\frac{z}{2}\right)} \quad \csc(z)=\frac{\cot^2\left(\frac{z}{2}\right)+1}{2\cot\left(\frac{z}{2}\right)}$$

Representations through Hyperbolic Functions

The cosecant function has representations using the hyperbolic functions;

$$\csc(z)=\frac{i}{\sinh(iz)} \quad \csc(z)=\frac{1}{\cosh\left(\frac{\pi i}{2}-iz\right)} \quad \csc(z)=\frac{i\left(1-\tanh^2\left(\frac{zi}{2}\right)\right)}{2\tanh\left(\frac{zi}{2}\right)}$$

$$\csc(z)=\frac{i\left(\coth^2\left(\frac{iz}{2}\right)-1\right)}{2\coth\left(\frac{iz}{2}\right)}$$

$$\csc(z)=i\,\operatorname{csc}h(iz) \quad \csc(iz)=-i\,\operatorname{csc}h(z) \quad \csc(z)=\operatorname{sec}h\left(\frac{\pi i}{2}-iz\right)$$

Applications

The cosecant function is used throughout mathematics, the exact sciences, and engineering.

SECANT

The secant function, denoted sec, is defined as the composite of the reciprocal function and the cosine function.

Explicitly, it is given as follows:

$$\sec x := \frac{1}{\cos x}$$

The domain of the function is the set of all $x \in \mathbb{R}$ for which $\cos x \neq 0$.

Key Data

Item	Value
Default domain	All real numbers except odd integer multiples of $\pi/2$.
Range	$(-\infty,-1]\cup[1,\infty)$, i.e., $\{y \mid \|y\| \geq 1\}$.
Period	2π, i.e., 360 degree.
Vertical asymptotes	Lines of the form $x = n\pi + \pi/2$. For n even, left hand limit is $+\infty$ and right hand limit is $-\infty$. For n odd, left hand limit is $-\infty$ and right hand limit is $+\infty$.
Local minimum values and points of attainment	Local minimum value of 1, attained at integer multiples of 2π.
Local maximum values and points of attainment	Local maximum value of -1, attained at odd integer multiples of π.
Important symmetries	Even function mirror symmetry about lines of the form $x = n\pi$, n varying over integers. half turn symmetry about points $(n\pi + \pi/2, 0)$, n varying over integers.
First derivative	$x \mapsto \sec x \tan x$.
Second derivative	$x \mapsto \sec x \tan^2 x + \sec^3 x$.
First antiderivative	$x \mapsto \ln\|\sec x + \tan x\| + C$.

Like the other trigonometric functions, the secant can be represented as a line segment associated with the unit circle. The diagram shows the secant for an angle of rotation θ of forty-five degrees (measured anti-clockwise from the positive x-axis). Line segment BF lies on the line that is tangent to the circle at point B. Line segment OF (shown in red) represents the secant, and is an extension of line segment OP. Remember that point P is a point on the circumference of the circle whose x and y coordinates represent the value of cos (θ) and sin (θ) respectively (the line segments representing the sine and cosine are also shown).

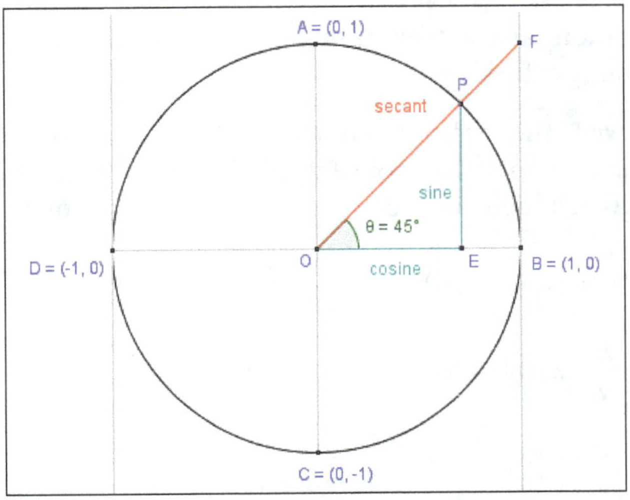

The secant as a line segment of the unit circle.

Here is a graphic of the secant function $f(x) = \sec(x)$ for real values of its argument x.

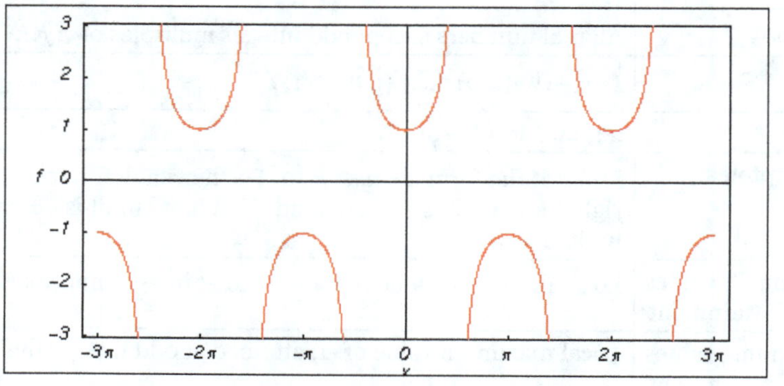

Representation through more General Functions

The secant function sec(z) can be represented using more general mathematical functions. As the ratio of one and the cosine function that is a particular case of the generalized hypergeometric, Bessel, Struve, and Mathieu functions, the secant function can also be represented as ratios of one and those special functions. Here are some examples:

$$\sec(z) = \frac{1}{{}_0F_1\left(;\frac{1}{2};-\frac{z^2}{4}\right)} \quad \sec(z) = \frac{1}{\sqrt{\frac{\pi z}{2}}J_{-\frac{1}{2}}(z)} \quad \sec(z) = \frac{1}{\sqrt{\frac{\pi i z}{2}}I_{-\frac{1}{2}}(iz)}$$

$$\sec(z) = -\frac{1}{\sqrt{\frac{\pi z}{2}}Y_{\frac{1}{2}}(z)} \quad \sec(z) = \frac{1}{1-\sqrt{\frac{\pi z}{2}}H_{\frac{1}{2}}(z)} \quad \sec(z) = \frac{1}{Ce(1,0,z)}$$

But these representations are not very useful because they include complicated special functions in the denominators.

It is more useful to write the secant function as particular cases of one special function. That can be done using doubly periodic Jacobi elliptic functions that degenerate into the secant function when their second parameter is equal to 0 or 1:

$$\sec(z) = dc(z\,|\,0) = nc(z\,|\,0) = ds\left(\frac{\pi}{2} - z\,|\,0\right)$$

$$= ns\left(\frac{\pi}{2} - z\,|\,0\right) = cn(iz\,|\,1) = dn(iz\,|\,1)$$

$$= ics\left(\frac{\pi i}{2} - iz\,|\,1\right) = ids\left(\frac{\pi i}{2} - iz\,|\,1\right)$$

Definition of the Secant Function for a Complex Argument

In the complex z-plane, the function sec(z) is defined using cos(z) or the exponential function e^w in the points iz and -iz through the formula:

$$\sec(z) = \frac{1}{\cos(z)} = \frac{2}{e^{iz} + e^{-iz}}$$

In the points $z = \pi / 2 + \pi k/; k \in \mathbb{Z}$, where cos(z) has zeros, the denominator of the last formula equals zero and sec(z) has singularities (poles of the first order).

Here are two graphics showing the real and imaginary parts of the secant function over the complex plane:

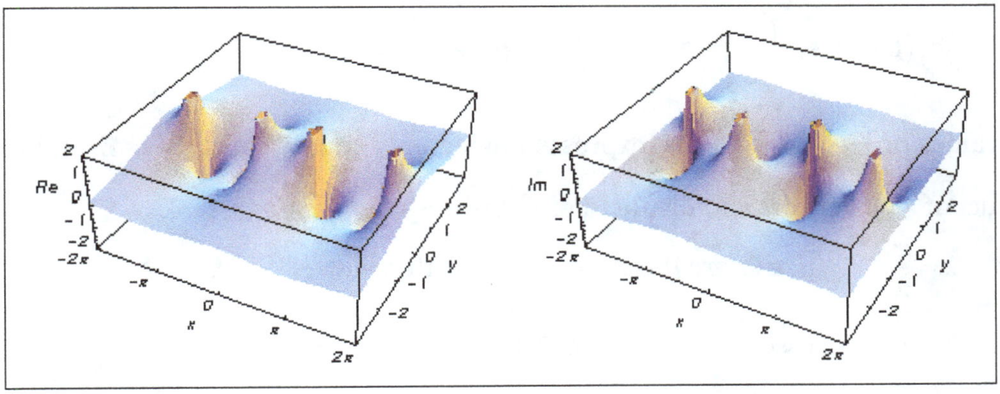

The Best-known Properties and Formulas for the Secant Function

Values in Points

Using the connection between the cosine and secant functions gives the following table of values of the secant function for angles between 0 and 2 π:

$$\sec(0) = 1 \quad \sec\left(\frac{\pi}{6}\right) = \frac{2}{\sqrt{3}} \quad \sec\left(\frac{\pi}{4}\right) = \sqrt{2} \quad \sec\left(\frac{\pi}{3}\right) = 2$$

$$\sec\left(\frac{\pi}{2}\right) = \overline{\infty} \quad \sec\left(\frac{2\pi}{3}\right) = -2 \quad \sec\left(\frac{3\pi}{4}\right) = -\sqrt{2} \quad \sec\left(\frac{5\pi}{6}\right) = -\frac{2}{\sqrt{3}}$$

$$\sec(\pi) = -1 \quad \sec\left(\frac{7\pi}{6}\right) = -\frac{2}{\sqrt{3}} \quad \sec\left(\frac{5\pi}{4}\right) = -\sqrt{2} \quad \sec\left(\frac{4\pi}{3}\right) = -2$$

$$\sec\left(\frac{3\pi}{2}\right)=\tilde{\infty}\ \sec\left(\frac{5\pi}{3}\right)=2\ \sec\left(\frac{7\pi}{4}\right)=\sqrt{2}\ \sec\left(\frac{11\pi}{6}\right)=\frac{2}{\sqrt{3}}$$

$$\sec(2\pi)=1\ \sec(\pi m)=(-1)^m\ /;m\in\mathbb{Z}\ \sec\left(\pi\left(\frac{1}{2}+m\right)\right)=\tilde{\infty}/;m\in\mathbb{Z}$$

General Characteristics

For real values of argument z, the values of sec(z) are real.

In the points $z=2\pi n/m/;n\in\mathbb{Z}\wedge m\in\mathbb{Z}$, the values of sec(z) are algebraic. In several cases they can be integers -2, -1, 1, or 2:

$$\sec(0)=1\ \sec\left(\frac{\pi}{3}\right)=2\ \sec\left(\frac{2\pi}{3}\right)=-2\ \sec(\pi)=-1$$

The values of $\sec\left(\frac{n\pi}{m}\right)$ can be expressed using only square roots if $n\in\mathbb{Z}$ and m is a product of a power of 2 and distinct Fermat primes {3, 5, 17, 257, etc}.

- $z=\frac{\pi}{2}+\pi k/;k\in\mathbb{Z}$ are the simple poles with residues $(-1)^{k-1}$.
- $z=\tilde{\infty}$ is an essential singular point.

It is a periodic function with a real period 2π:

$$\sec(z+2\pi)=\sec(z)$$
$$\sec(z)=\sec(z+2\pi k)/;k\in\mathbb{Z}$$
$$\sec(z)=(-1)^k\sec(z+\pi k)/;k\in\mathbb{Z}$$

The function sec(z) is an even function with mirror symmetry:

$$\sec(z)=\sec(z)\ \sec(z)=\overline{\sec(z)}$$

Differentiation

The first derivative of sec(z) has simple representations using either the tan(z) function or the sec(z) function:

$$\frac{\partial\sec(z)}{\partial z}=\tan(z)\sec(z)$$

The nth derivative of sec(z) has much more complicated representations.

Than symbolic n[th] derivatives for sin(z) and cos(z):

$$\frac{\partial \sec(z)}{\partial z} = \sec(z)\left(\delta_n + (n+1)!\sum_{k=0}^{n}\sum_{j=0}^{\left\lceil \frac{k-1}{2} \right\rceil}\frac{(-1)^k 2^{1-k}(k-2j)^n \sec^k(z)}{(k+1)j!(k-j)!(n-k)!}\right.$$

$$\left.\cos\left(\frac{\pi n}{2} + (k-2j)z\right)\right) /; n \in \mathbb{Z}$$

where δ_n is the Kronecker delta symbol: $\delta_0 = 1$ and $\delta_n = 0/; n \neq 0$.

Ordinary Differential Equation

The function sec(z) satisfies the following first-order nonlinear differential equation:

$$w'(z)^2 - w(z)^4 + w(z)^2 = 0/; w(z) = \sec(z)$$

Series Representation

The function sec(z) has the following series expansion at the origin that converges for all finite values z with $|z| < \frac{\pi}{2}$:

$$\sec(z) = 1 + \frac{z^2}{2} + \frac{5z^4}{24} + \ldots = \sum_{k=0}^{\infty}\frac{(-1)^k E_{2k} z^{2k}}{(2k)!},$$

Where E_{2k} are the Euler numbers.

The secant function sec(z) can also be presented using other kinds of series by the following formulas:

$$\sec(z) = \pi\sum_{k=0}^{\infty}\frac{(-1)^k(2k+1)}{\left(k+\frac{1}{2}\right)^2 \pi^2 - z^2} /; \frac{z}{\pi} - \frac{1}{2} \notin \mathbb{Z}$$

$$\sec^2(z) = \sum_{k=-\infty}^{\infty}\frac{1}{\left(z+\pi\left(k+\frac{1}{2}\right)\right)^2} /; \frac{z}{\pi} - \frac{1}{2} \notin \mathbb{Z}$$

Integral Representation

The function sec(z) has a well-known integral representation through the following definite integral along the positive part of the real axis:

$$\sec(z) = \frac{2}{\pi}\int_0^{\infty}\frac{t^{\frac{2z}{\pi}}}{t^2+1}dt /; |\mathrm{Re}(z)| < \frac{\pi}{2}$$

Product Representation

The famous infinite product representation for cos(z) can be easily rewritten as the following product representation for the secant function:

$$\sec(z) = \prod_{k=1}^{\infty} \frac{\pi^2 (2k-1)^2}{\pi^2 (2k-1)^2 - 4z^2}$$

Limit Representation

The secant function has the following limit representation:

$$\sec(z) = \lim_{n\to\infty} \sum_{k=-n}^{n} \frac{(-1)^k}{\left(k+\frac{1}{2}\right) - z} /; \frac{z}{\pi} - \frac{1}{2} \notin \mathbb{Z}$$

Indefinite Integration

Indefinite integrals of expressions involving the secant function can sometimes be expressed using elementary functions. However, special functions are frequently needed to express the results even when the integrands have a simple form (if they can be evaluated in closed form). Here are some examples:

$$\int \sec(z)dz = \log\left(\cot\left(\frac{z}{2} - \frac{\pi}{4}\right)\right)$$

$$\int \sqrt{\sec(z)}dz = 2\cos^{\frac{1}{2}}(z) F\left(\frac{z}{2}\middle| 2\right) \sec^{\frac{1}{2}}(z)$$

$$\int \sec^{v}(az)dz = \frac{\csc(az)\sec^{v-1}(az)\sqrt{\sin^2(az)}}{av-a} {}_2F_1\left(\frac{1-v}{2}, \frac{1}{2}; \frac{3-v}{2}; \cos^2(az)\right)$$

Definite Integration

Definite integrals that contain the secant function are sometimes simple and their values can be expressed through elementary functions. Here is one example:

$$\int_0^{\frac{\pi}{4}} t\sec^2(t)dt = \frac{1}{4}(\pi - \log(4))$$

Some special functions can be used to evaluate more complicated definite integrals.

For example, polygamma and gamma functions and the Catalan constant are needed to express the following integrals:

$$\int_0^{\frac{\pi}{4}} t\sec(t)dt = \frac{1}{32}\left(\begin{array}{l} 8\pi\left(\log\left(1-(-1)^{3/4}\right)-\log\left(1+(-1)^{3/4}\right)\right) \\ +\sqrt[4]{-1}\left(\psi^{(1)}\left(\frac{1}{8}\right)-i\psi^{(1)}\left(\frac{3}{8}\right)-\psi^{(1)}\left(\frac{5}{8}\right)+i\psi^{(1)}\left(\frac{7}{8}\right)\right)-64C \end{array}\right)$$

$$\int_0^{\frac{\pi}{2}} \sec^a(t)dt = \frac{\sqrt{\pi}}{2\Gamma\left(1-\frac{a}{2}\right)}\Gamma\left(\frac{1-a}{2}\right)/;\mathrm{Re}(a)<1$$

Finite Summation

Finite sums that contain the secant function have the following simple values:

$$\sum_{k=0}^{n}\sec\left(\frac{2k\pi}{2n+1}\right)=\frac{1}{2}(-1)^n(2n+1)+\frac{1}{2}/;n\in\mathbb{N}$$

$$\sum_{k=1}^{\left\lfloor\frac{n-1}{2}\right\rfloor}\sec^2\left(\frac{k\pi}{n}\right)=\frac{1}{12}\left(4n^2-(-1)^n\left(2n^2+1\right)-7\right)/;n\in\mathbb{N}^+$$

$$\sum_{k=0}^{n-1}\sec^2\left(\frac{\pi k}{n}+z\right)=n^2\csc^2\left(\frac{1}{2}n(2z+\pi)\right)/;n\in\mathbb{N}^+$$

$$\sum_{k=0}^{n}\sec^2\left(\frac{2\pi k}{n}+z\right)=\frac{1}{2}\left(1-(-1)^n\right)n^2\csc^2\left(\frac{1}{2}n(2z+\pi)\right)$$
$$+\frac{1}{4}\left(1+(-1)^n\right)n^2\csc^2\left(\frac{1}{4}n(2z+\pi)\right)/;n\in\mathbb{N}^+$$

$$\sum_{k=1}^{n}\frac{1}{2^2k}\sec^2\left(\frac{z}{2^k}\right)=\csc^2(z)-\frac{1}{2^2n}\csc^2\left(\frac{z}{2^n}\right)/;n\in\mathbb{N}.$$

Infinite Summation

For $n->\infty$ gives the following value for the corresponding infinite sum:

$$\sum_{k=1}^{\infty}\frac{1}{2^2k}\sec^2\left(\frac{z}{2^k}\right)=\csc^2(z)-\frac{1}{z^2}.$$

Finite Products

The following finite product from the secant can be represented through the cosecant function:

$$\prod_{k=0}^{n-1}\sec\left(z+\frac{\pi k}{n}\right)=(-1)^{n-1}2^{n-1}\csc\left(\frac{n(\pi-2z)}{2}\right)/;n\in\mathbb{N}^+.$$

Infinite Products

The following infinite product from the secant can be represented through the cosecant function:

$$\prod_{k=1}^{\infty} \sec\left(\frac{z}{2^k}\right) = z\csc(z)$$

Addition Formulas

The secants of a sum and a difference can be represented by the following formulas that are derived from the cosines of a sum and a difference:

$$\sec(a+b) = \frac{1}{\cos(b)\cos(a) - \sin(a)\sin(b)}$$

$$\sec(a-b) = \frac{1}{\cos(b)\cos(a) + \sin(a)\sin(b)}$$

Multiple Arguments

In the case of multiple arguments z, 2z, 3z, etc., the function $\sec(z)$ can be represented as a rational function including powers of a secant. Here are two examples:

$$\sec(2z) = \frac{\sec^2(z)}{2 - \sec^2(z)}$$

$$\sec(3z) = \frac{\sec^3(z)}{4 - 3\sec^2(z)}$$

Half-angle Formulas

The secant of a half-angle can be represented by the following simple formula that is valid in a vertical strip:

$$\sec\left(\frac{z}{2}\right) = \frac{\sqrt{2}}{\sqrt{1+\cos(z)}} /; |\operatorname{Re}(z)| < \pi \vee \operatorname{Re}(z) = -\pi \wedge \operatorname{Im}(z) > 0 \vee \operatorname{Re}(z) = \pi \wedge \operatorname{Im}(z) < 0.$$

To make this formula correct for all complex , z a complicated prefactor is needed:

$$\sec\left(\frac{z}{2}\right) = \frac{c(z)}{\sqrt{1+\cos(z)}} /; c(z) = (-1)^{\left\lfloor \frac{\operatorname{Re}(z)+\pi}{2\pi} \right\rfloor} \sqrt{2}\left(1 - \left(1 + (-1)^{\left\lfloor \frac{\operatorname{Re}(z)+\pi}{2\pi} \right\rfloor + \left\lfloor \frac{\operatorname{Re}(z)+\pi}{2\pi} \right\rfloor}\right)\theta(-\operatorname{Im}(z))\right),$$

where $c(z)$ contains the unit step, real part, imaginary part, and the floor functions.

Sums of Two Direct Functions

The sum and difference of two secant functions can be described by the following formulas:

$$\sec(a)+\sec(b)=2\cos\left(\frac{a}{2}-\frac{b}{2}\right)\cos\left(\frac{a}{2}+\frac{b}{2}\right)\sec(a)\sec(b)$$

$$\sec(a)-\sec(b)=2\sin\left(\frac{a}{2}-\frac{b}{2}\right)\sin\left(\frac{a}{2}+\frac{b}{2}\right)\sec(a)\sec(b)$$

Products involving the Direct Function

The product of two secants and the product of a secant and a cosecant have the following representations:

$$\sec(a)\sec(b)=\frac{2}{\cos(a-b)+\cos(a+b)}$$

$$\sec(a)\csc(b)=\frac{2}{\sin(a+b)+\sin(a-b)}$$

Inequalities

One of the most famous inequalities for a secant function is the following:

$$\sec(x)>x\csc(x)/;0<x<\frac{\pi}{2}\wedge x\in\mathbb{R}.$$

Relations with its Inverse Function

There are simple relations between the function $\sec(z)$ and its inverse function $\sec^{-1}(z)$:

$$\sec\left(\sec^{-1}(z)\right)=z \quad \sec^{-1}\left(\sec(z)\right)=z/;0<\operatorname{Re}(z)<\pi\vee\operatorname{Re}(z)$$
$$=0\wedge\operatorname{Im}(z)\geq 0\vee\operatorname{Re}(z)=\pi\wedge\operatorname{Im}(z)\leq 0.$$

The second formula is valid at least in the vertical strip $0<\operatorname{Re}(z)<\pi$. Outside of this strip a much more complicated relation (containing the unit step, real part, and the floor functions) holds:

$$\sec\left(\sec^{-1}(z)\right)=\frac{\pi}{2}\left(1-(-1)^{-\left\lfloor\frac{\operatorname{Re}(z)}{\pi}\right\rfloor}\right)+(-1)^{\left\lfloor-\frac{\operatorname{Re}(z)}{\pi}\right\rfloor}\left(\left(1+(-1)^{\left\lfloor\frac{\operatorname{Re}(z)}{\pi}\right\rfloor+\left\lfloor-\frac{\operatorname{Re}(z)}{\pi}\right\rfloor}\right)\theta(\operatorname{Im}(z))-1\right)$$
$$\left(z+\pi\left\lfloor\frac{\operatorname{Re}(z)}{\pi}\right\rfloor\right).$$

Representations through other Trigonometric Functions

Secant and cosecant functions are connected by a very simple formula that contains the linear function in the argument:

$$\sec(z) = \csc\left(\frac{\pi}{2} - z\right).$$

The secant function can also be represented using other trigonometric functions by the following formulas:

$$\sec(z) = \frac{1}{\sin\left(\frac{\pi}{2} - z\right)} \quad \sec(z) = \frac{1 + \tan^2\left(\frac{z}{2}\right)}{1 - \tan^2\left(\frac{z}{2}\right)} \quad \sec(z) = \frac{\cot^2\left(\frac{z}{2}\right) + 1}{\cot^2\left(\frac{z}{2}\right) - 1}$$

Representations through Hyperbolic Functions

The secant function has representations using the hyperbolic functions:

$$\sec(z) = \frac{i}{\sinh\left(\frac{\pi i}{2} - iz\right)} \quad \sec(z) = \frac{1}{\cosh(iz)} \quad \sec(z) = \frac{1 - \tanh^2\left(\frac{iz}{2}\right)}{1 + \tanh^2\left(\frac{iz}{2}\right)}$$

$$\sec(z) = \frac{\coth^2\left(\frac{iz}{2}\right) - 1}{\coth^2\left(\frac{iz}{2}\right) + 1}$$

$$\sec(z) = i\,\mathrm{csch}\left(\frac{\pi i}{2} - iz\right) \quad \sec(z) = \mathrm{sech}(i,z) \quad \sec(iz) = \mathrm{sech}(z)$$

Applications

The secant function is used throughout mathematics, the exact sciences, and engineering.

COTANGENT

The cotangent function, denoted cot, is defined as the pointwise quotient of the cosine function by the sine function:

$$\cot x := \frac{\cos x}{\sin x}$$

It can be defined as the composite of the reciprocal function and the tangent function, with the caveat that the cotangent function is defined to be zero at all points where the tangent function is undefined.

Integration

First Antiderivative: Using f'/f Formulation

We use the integration form integration of quotient of derivative of function by function:

$$\int \frac{f'(x)}{f(x)}dx = \ln|f(x)| + C$$

In our case, we write:

$$\int \cot x\, dx = \int \frac{\cos x}{\sin x}dx$$

This is the f'/f form where $f = sin$, so we get:

$$\ln|\sin x| + C$$

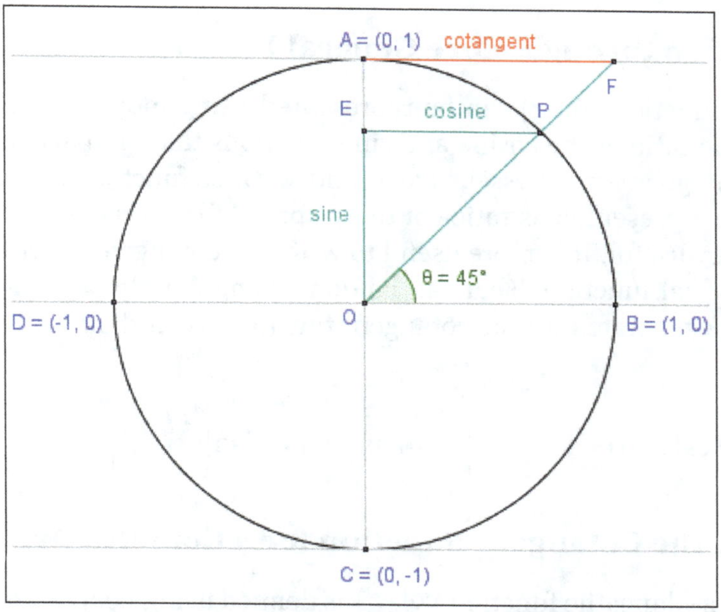

The cotangent as a line segment of the unit circle.

Like the other trigonometric functions, the cotangent can be represented as a line segment associated with the unit circle. The diagram shows the cotangent for an angle of rotation θ of forty-five degrees (measured anti-clockwise from the positive x-axis). Line segment AF (shown in red) is the cotangent, and lies on the line that is tangent to the circle at point A. Line segment PF is an extension of line segment OP (and, incidentally, also happens to be the secant). Remember that point P is a point on the circumference of the unit circle whose x and y coordinates represent the value of cos (θ) and sin (θ) respectively (the line segments representing the sine and cosine are also shown). Remember that one definition of the tangent function is as the quotient of the sine and cosine functions.

Here is a graphic of the cotangent function $f(x)=\cot(x)$ for real values of its argument x.

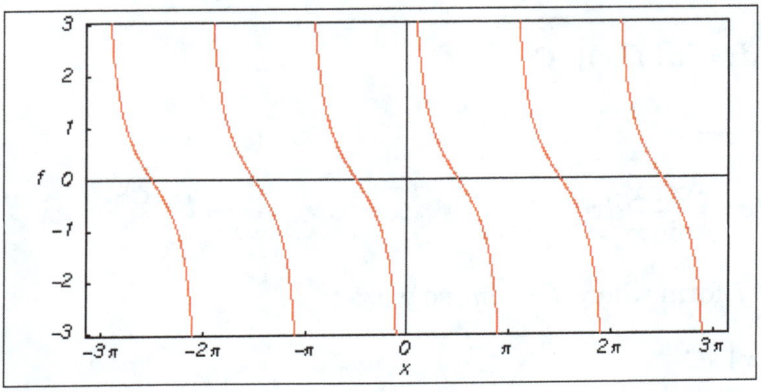

Representation through more General Functions

The cotangent function $\cot(z)$ can be represented using more general mathematical functions. As the ratio of the cosine and sine functions that are particular cases of the generalized hypergeometric, Bessel, Struve, and Mathieu functions, the cotangent function can also be represented as ratios of those special functions. But these representations are not very useful. It is more useful to write the cotangent function as particular cases of one special function. That can be done using doubly periodic Jacobi elliptic functions that degenerate into the cotangent function when their second parameter is equal to 0 or 1.

$$\cot(z)=cs(z\,|\,0)=\operatorname{sc}\left(\frac{\pi}{2}-z\middle|\,0\right)=i\operatorname{ns}(iz\,|\,1)=i\operatorname{sn}\left(\frac{\pi i}{2}-iz\middle|\,1\right).$$

Definition of the Cotangent Function for a Complex Argument

In the complex z-plane, the function $\cot(z)$ is defined using $\cos(z)$ and $\sin(z)$ or the exponential function e^{w} in the points iz and iz through the formula:

$$\cot(z)=\frac{\cos(z)}{\sin(z)}=\frac{i\left(e^{iz}+e^{-1z}\right)}{e^{iz}-e^{-1z}}.$$

In the points $z=\pi k/;\in\mathbb{Z}$, where $\sin(z)$ has zeros, the denominator of the last formula equals zero and $\cot(z)$ has singularities (poles of the first order).

Here are two graphics showing the real and imaginary parts of the cotangent function over the complex plane.

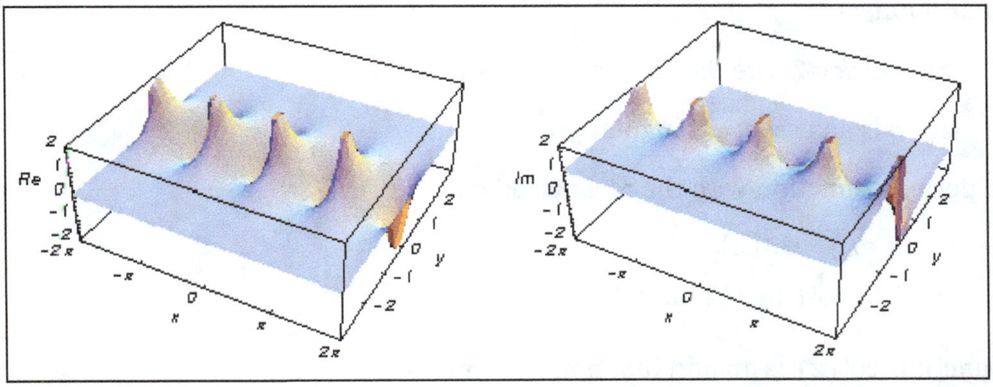

The Best-known Properties and Formulas for the Cotangent Function

Values in Points

Students usually learn the following basic table of values of the cotangent function for special points of the circle:

$$\cot(0)=\overline{\infty} \quad \cot\left(\frac{\pi}{6}\right)=\sqrt{3} \quad \cot\left(\frac{\pi}{4}\right)=1 \quad \cot\left(\frac{\pi}{3}\right)=\frac{1}{\sqrt{3}}$$

$$\cot\left(\frac{\pi}{2}\right)=0 \quad \cot\left(\frac{2\pi}{3}\right)=-\frac{1}{\sqrt{3}} \quad \cot\left(\frac{3\pi}{4}\right)=-1 \quad \cot\left(\frac{5\pi}{6}\right)=-\sqrt{3}$$

$$\cot(\pi)=\overline{\infty}$$

$$\cot(\pi m)=\overline{\infty}/;m\in\mathbb{Z} \quad \cot\left(\pi\left(\frac{1}{2}+m\right)\right)=0/;m\in\mathbb{Z}$$

General Characteristics

For real values of argument z, the values of $\cot(z)$ are real.

In the points $z=\pi n/m/;n\in\mathbb{Z}\wedge m\in\mathbb{Z}$, the values of $\cot(z)$ are algebraic. In several cases they can be integers -1, 0, or 1:

$$\cot\left(-\frac{\pi}{4}\right)=-1 \quad \cot\left(\frac{\pi}{2}\right)=0 \quad \cot\left(\frac{\pi}{4}\right)=1.$$

The values of $\cot\left(\frac{n\pi}{m}\right)$ can be expressed using only square roots if $n\in\mathbb{Z}$ and m is a product of a power of 2 and distinct Fermat primes {3, 5, 17, 257, etc}.

The function $\cot(z)$ is an analytical function of z that is defined over the whole complex z-plane and does not have branch cuts and branch points. It has an infinite set of singular points:

- $z = \pi k /; k \in \mathbb{Z}$ are the simple poles with residues -1.
- $z = \bar{\infty}$ is an essential singular point.

It is a periodic function with the real period π:

$$\cot(z+\pi) = \cot(z)$$
$$\cot(z) = \cot(z+\pi k) /; k \in \mathbb{Z}$$

The function $\cot(z)$ is an odd function with mirror symmetry:

$$\cot(-z) = -\cot(z) \quad \cot(\bar{z}) = \overline{\cot(z)}$$

Differentiation

The first derivative of $\cot(z)$ has simple representations using either the $\sin(z)$ function or the $\csc(z)$ function:

$$\frac{\partial \cot(z)}{\partial z} = -\frac{1}{\sinh^2(z)} = -\csc^2(z).$$

The nth derivative of $\cot(z)$ has much more complicated representations than symbolic n^{th} derivatives for $\sin(z)$ and $\cos(z)$:

$$\frac{\partial^n \cot(z)}{\partial z^n} = \cot(z)\delta_n - \delta_{n-1}\csc^2(z) - n\sum_{k=0}^{n-1}\sum_{j=0}^{k-1}\frac{(-1)^j \sin^{-2k-2}(z) 2^{n-2k}(k-j)^{n-1}}{k+1}$$
$$\binom{n-1}{k}\binom{2k}{j}\sin\left(\frac{\pi n}{2} + 2(k-j)z\right) /; n \in \mathbb{N},$$

where δ_n is the Kronecker delta symbol: $\delta_0 = 1$ and $\delta_n = 0 /; n \neq 0$.

Ordinary Differential Equation

The function $\cot(z)$ satisfies the following first-order nonlinear differential equation:

$$w'(z) + w(z)^2 + 1 = 0 /; w(z) = \cot(z) \wedge w\left(\frac{\pi}{2}\right) = 0$$

Series Representation

The function $\cot(z)$ has a simple Laurent series expansion at the origin that converges for all finite values z with $0<|z|<\pi$:

$$\cot(z)=\frac{1}{z}-\frac{z}{3}-\frac{z^3}{45}-\ldots\frac{1}{z}+\sum_{k=1}^{\infty}\frac{(-1)^k 2^{2k} B_{2k} z^{2k-1}}{(2k)!}$$

where B_{2k} are the Bernoulli numbers.

Integral Representation

The function $\cot(z)$ has a well-known integral representation through the following definite integral along the positive part of the real axis:

$$\cot(z)=\frac{2}{\pi}\int_0^{\infty}\frac{t^{1-\frac{2z}{\pi}}-1}{t^2-1}dt/;0<\mathrm{Re}(z)<\frac{\pi}{2}$$

Continued Fraction Representations

The function $\cot(z)$ has the following simple continued fraction representation:

$$\cot(z)=\frac{1}{z}-\cfrac{4\pi^{-2}z}{1+\cfrac{1-4\pi^{-2}z^2}{3+\cfrac{4(4-4\pi^{-2}z^2)}{5+\cfrac{9(9-4\pi^{-2}z^2)}{7+\cfrac{16(16-4\pi^{-2}z^2)}{9+\cfrac{25(25-4\pi^{-2}z^2)}{11+\cfrac{36(36-4\pi^{-2}z^2)}{13+\ldots}}}}}}}$$

Indefinite Integration

Indefinite integrals of expressions that contain the cotangent function can sometimes be expressed using elementary functions. However, special functions are frequently needed to express the results even when the integrands have a simple form (if they can be evaluated in closed form). Here are some examples:

$$\int\cot(z)=\log(\sin(z))$$

$$\int\sqrt{\cot(z)}\,dz=\frac{1}{2\sqrt{2}}\left(\begin{array}{c}-2\tan^{-1}\left(\sqrt{2}\cot^{\frac{1}{2}}(z)+1\right)+2\tan^{1}\left(1-\sqrt{2}\cot^{\frac{1}{2}}(z)\right)\\-\log\left(-\cot(z)+\sqrt{2}\cot^{\frac{1}{2}}(z)-1\right)+\log\left(\cot(z)+\sqrt{2}\cot^{\frac{1}{2}}(z)+1\right)\end{array}\right)$$

$$\int\cot^{v}(az)dz=-\frac{\cot^{v+1}(az)}{va+a}{}_2F_1\left(\frac{v+1}{2},1;\frac{v+3}{2};-\cot^2(az)\right).$$

Definite Integration

Definite integrals that contain the cotangent function are sometimes simple. For example, the famous Catalan constant can be defined as the value of the following integral:

$$\int_0^{\frac{\pi}{4}}\log(\cot(t))dt=C$$

This constant also appears in the following integral:

$$\int_0^{\frac{\pi}{4}}t\cot(t)dt=\frac{1}{16}\left(8C+i\pi^2+4\pi\log(1-i)\right)$$

Some special functions can be used to evaluate more complicated definite integrals. For example, to express the following integral, the Gauss hypergeometric function is needed:

$$\int_0^{\frac{\pi}{4}}\sin^{\alpha-1}(t)\cot^{\beta}(t)dt=\frac{2^{\frac{1-\alpha}{2}}}{\alpha-\beta}{}_2F_1\left(1,\frac{1-\beta}{2};\frac{1}{2}(\alpha-\beta+2);-1\right)/;\mathrm{Re}(\alpha-\beta)>0$$

Finite Summation

The following finite sum that contains a cotangent function can be expressed in terms of a cotangent function:

$$\sum_{k=0}^{n-1}\cot^2\left(\frac{\pi k}{n}+z\right)=\cot^2(nz)n^2+n^2-n/;n\in\mathbb{N}^+$$

Other finite sums that contain a cotangent function can be expressed in terms of a polynomial function:

$$\sum_{k=1}^{n-1}\cot^2\left(\frac{\pi k}{n}\right)=\frac{(n-1)(n-2)}{3}/;n\in\mathbb{N}^+$$

$$\sum_{k=1}^{\left\lfloor\frac{n-1}{2}\right\rfloor}\cot^2\left(\frac{k\pi}{n}\right)=\frac{1}{6}(n-1)(n-2)/;n\in\mathbb{N}^+$$

$$\sum_{k=0}^{n-1} (-1)^k \cot\left(\frac{(2k+1)\pi}{4n}\right) = n /; n \in \mathbb{N}^+$$

$$\sum_{k=0}^{2n-1} (-1)^k \cot\left(\frac{(2k+1)\pi}{8n}\right) = 2n /; n \in \mathbb{N}^+$$

$$\sum_{k=1}^{n} \cot^4\left(\frac{k\pi}{2n+1}\right) = \frac{1}{45} n(2n-1)\left(4n^2+10n-9\right) /; n \in \mathbb{N}^+$$

Infinite Summation

The following infinite sum that contains the cotangent function has a very simple value:

$$\sum_{k=1}^{\infty} \frac{1}{k^3} \cot\left(\frac{1}{2} k\pi\left(1+\sqrt{5}\right)\right) = -\frac{\pi^3}{45\sqrt{5}}.$$

Finite Products

The following finite product from the cotangent has a very simple value:

$$\prod_{k=1}^{n-1} \cot\left(\frac{k\pi}{n}\right) = -\frac{(-1)^n}{n} \sin\left(\frac{n\pi}{2}\right) /; n \in \mathbb{N}^+$$

Addition Formulas

The cotangent of a sum can be represented by the rule: "the cotangent of a sum is equal to the product of the cotangents minus one divided by a sum of the cotangents." A similar rule is valid for the cotangent of the difference:

$$\cot(a+b) = \frac{\cot(a)\cot(b)-1}{\cot(a)+\cot(b)}$$

$$\cot(a-b) = \frac{\cot(a)\cot(b)+1}{\cot(b)-\cot(a)}.$$

Multiple Arguments

In the case of multiple arguments z, 2z, 3z, etc., the function $\cot(z)$ can be represented as the ratio of the finite sums that contains powers of cotangents:

$$\cot(2z) = \frac{\cot^2(z)-1}{2\cot(z)} = \frac{1}{2}\left(\cot(z)-\tan(z)\right)$$

$$\cot(3z) = \frac{\cot^3(z) - 3\cot(z)}{3\cot^2(z) - 1}$$

$$\cot(nz) = \frac{(-1)^{n-1}}{\cot^{(-1)^n}(z)\sum_{k=0}^{\left\lfloor\frac{n-1}{2}\right\rfloor}(-1)^k\binom{n}{2\left\lfloor\frac{n-1}{2}\right\rfloor - 2k + 1}\cot^{2k}(z)}$$

$$\sum_{k=0}^{\left\lfloor\frac{n}{2}\right\rfloor}(-1)^k\binom{n}{2\left\lfloor\frac{n}{2}\right\rfloor - 2k}\cot^{2k}(z) /; n \in \mathbb{N}^+.$$

Half-angle Formulas

The cotangent of a half-angle can be represented using two trigonometric functions by the following simple formulas:

$$\cot\left(\frac{z}{2}\right) = \cot(z) + \csc(z)$$

$$\cot\left(\frac{z}{2}\right) = \frac{\sin(z)}{1 - \cos(z)}.$$

The sine function in the last formula can be replaced by the cosine function. But it leads to a more complicated representation that is valid in some vertical strip:

$$\cot\left(\frac{z}{2}\right) = \sqrt{\frac{1+\cos(z)}{1-\cos(z)}} /; 0 < \operatorname{Re}(z) < \pi \vee \operatorname{Re}(z) = 0 \wedge \operatorname{Im}(z) < 0 \vee \operatorname{Re}(z) = \pi \wedge \operatorname{Im}(z) \le 0.$$

To make this formula correct for all complex z, a complicated prefactor is needed:

$$\cot\left(\frac{z}{2}\right) = \sqrt{\frac{1+\cos(z)}{1-\cos(z)}} /; c(z) = (-1)^{\left\lfloor\frac{2\operatorname{Re}(z)-\pi}{2\pi}\right\rfloor}\left(1 - \left(1 + (-1)^{\left\lfloor\frac{\operatorname{Re}(z)}{\pi}\right\rfloor + \left\lfloor-\frac{\operatorname{Re}(z)}{\pi}\right\rfloor}\right)\theta(\operatorname{Im}(z))\right),$$

where $c(z)$ contains the unit step, real part, imaginary part, the floor, and the round functions.

Sums of Two Direct Functions

The sum of two cotangent functions can be described by the rule: "the sum of cotangents is equal to the sine of the sum multiplied by the cosecants." A similar rule is valid for the difference of two cotangents:

$$\cot(a)+\cot(b)=\csc(a)\csc(b)\sin(a+b)$$
$$\cot(a)-\cot(b)=-\csc(a)\csc(b)\sin(a-b).$$

Products involving the Direct Function

The product of two cotangents and the product of the cotangent and tangent have the following representations:

$$\cot(a)\cot(b)=\frac{\cos(a-b)+\cos(a+b)}{\cos(a-b)-\cos(a+b)}$$
$$\cot(a)\tan(b)=\frac{\sin(a+b)-\sin(a-b)}{\sin(a-b)+\sin(a+b)}$$

Inequalities

The most famous inequality for the cotangent function is the following:

$$\cot(x)<\frac{1}{x}/;0<x<\pi\wedge x\in\mathbb{R}.$$

Relations with its Inverse Function

There are simple relations between the function $\cot(z)$ and its inverse function $\cot^{-1}(z)$:

$$\cot\left(\cot^{-1}(z)\right)=z \quad \cot^{-1}(\cot(z))=z/;|\mathrm{Re}(z)|<\frac{\pi}{2}\vee\mathrm{Re}(z)$$
$$=-\frac{\pi}{2}\wedge\mathrm{Im}(z)<0\vee\mathrm{Re}(z)=\frac{\pi}{2}\wedge\mathrm{Im}(z)\geq 0.$$

The second formula is valid at least in the vertical strip $-\frac{\pi}{2}<\mathrm{Re}(z)<\frac{\pi}{2}$. Outside of this strip a much more complicated relation (that contains the unit step, real part, and the floor functions) holds:

$$\cot^{-1}(\cot(z))=z+\pi\left\lfloor\frac{1}{2}-\frac{\mathrm{Re}(z)}{\pi}\right\rfloor-\frac{1}{2}\pi\left(1+(-1)^{\left\lfloor\frac{\mathrm{Re}(z)}{\pi}+\frac{1}{2}\right\rfloor+\left\lfloor-\frac{\mathrm{Re}(z)}{\pi}-\frac{1}{2}\right\rfloor}\right)$$
$$\theta(-\mathrm{Im}(z))/;\frac{z}{\pi}-\frac{1}{2}\notin\mathbb{Z}$$

Representations through other Trigonometric Functions

Cotangent and tangent functions are connected by a very simple formula that contains the linear function in the following argument:

$$\cot(z) = \tan\left(\frac{\pi}{2} - z\right).$$

The cotangent function can also be represented using other trigonometric functions by the following formulas:

$$\cot(z) = \frac{\sin\left(\frac{\pi}{2} - z\right)}{\sin(z)} \quad \cot(z) = \frac{\cos(z)}{\cos\left(\frac{\pi}{2} - z\right)}$$

$$\cot(z) = \frac{\csc(z)}{\csc\left(\frac{\pi}{2} - z\right)} \quad \cot(z) = \frac{\sec\left(\frac{\pi}{2} - z\right)}{\sec(z)}$$

Representations through Hyperbolic Functions

The cotangent function has representations using the hyperbolic functions:

$$\cot(z) = \frac{\sinh\left(\frac{i\pi}{2} - iz\right)}{\sinh(iz)} \quad \cot(z) = \frac{\cosh(iz)}{\cosh\left(\frac{i\pi}{2} - iz\right)} \quad \cot(z) = -i\tanh\left(\frac{\pi i}{2} - iz\right) \cot(z) = i\coth(iz)$$

$$\coth(iz) = -i\coth(z) \quad \cot(z) = \frac{\operatorname{csch}(iz)}{\operatorname{csch}\left(\frac{\pi i}{2} - iz\right)} \quad \cot(z) = \frac{\operatorname{sech}\left(\frac{\pi i}{2} - iz\right)}{\operatorname{sech}(iz)}.$$

Applications

The cotangent function is used throughout mathematics, the exact sciences, and engineering.

References

- Trigonometric-function, science: britannica.com, Retrieved 04 August, 2020
- Trigonometric-functions-definition-examples-quiz, academy-lesson: study.com, Retrieved 28 June, 2020
- Cosine-function: varsitytutors.com, Retrieved 16 May, 2020

- Tangent: mathworld.wolfram.com, Retrieved 28 August, 2020
- Cosecant-function: calculus.subwiki.org, Retrieved 05 April, 2020
- Secant-function, mathematics-trigonometry: technologyuk.net, Retrieved 25 March, 2020
- Cotangent-function: calculus.subwiki.org, Retrieved 17 February, 2020

3

Understanding Inverse Trigonometric Functions

Inverse trigonometric functions are the inverse functions of sine, cosine, tangent, cotangent, secant, and cosecant functions. They can be used for obtaining an angle from any of the angle's trigonometric ratios. All the concepts related to inverse trigonometric functions have been carefully analyzed in this chapter.

BASIC CONCEPTS OF INVERSE TRIGONOMETRIC FUNCTIONS

The inverse trigonometric functions are arcus functions or anti trigonometric functions. These are the inverse functions of the trigonometric functions with suitably restricted domains.

If y=f(x) and x=g(y) are two functions such that f (g(y)) = y and g (f(y)) = x, then f and y are said to be inverse of each other i.e.,

$$g = f^{-1}$$

$$\text{If } y = f(x) \text{ then } x = f^{-1}(y)$$

Inverse Trigonometric Formulas

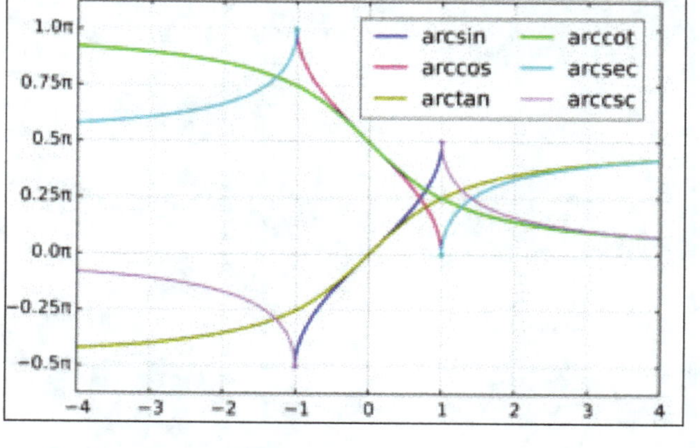

The inverse trigonometric functions are the inverse functions of the trigonometric functions written as $\cos^{-1} x$, $\sin^{-1} x$, $\tan^{-1} x$, $\cot^{-1} x$, $\mathrm{cosec}^{-1} x$, $\sec^{-1} x$.

The inverse trigonometric functions are multi-valued. For example, there are multiple values of ω such that $z = \sin\omega$, so $\sin^{-1} z$ is not uniquely defined unless a principal value is defined. Such principal values are sometimes denoted with a capital letter so, for example, the principal value of the inverse sine may be variously denoted $\sin^{-1} z$ or arc (sinz).

Let's say, if $y = \sin x$, then $x = \sin^{-1} y$, similarly for other trigonometric functions. This is one of the inverse trigonometric formulas. Now, $y = \sin^{-1} (x)$, $y \in [\pi/2 , \pi/2]$ and $x \in [-1,1]$.

- Thus, $\sin^{-1} x$ has infinitely many values for given $x \in [-1, 1]$.
- There is only one value among these values which lies in the interval $[\pi/2, \pi/2]$. This value is called the principal value.

Domain and Range of Inverse Trigonometric Formulas

Function	Domain	Range
$\sin^{-1}x$	[-1,1]	$[-\pi/2,\pi/2]$
$\cos^{-1}x$	[-1,1]	$[0,\pi]$
$\tan^{-1}x$	R	$(-\pi/2,\pi/2)$
$\cot^{-1}x$	R	$(0,\pi)$
$\sec^{-1}x$	R-(-1,1)	$[0,\pi]-\{\pi/2\}$
$\mathrm{cosec}^{-1}x$	R-(-1,1)	$[-\pi/2,\pi/2]-\{0\}$

Example: Find the exact value of each expression without a calculator, in $[0,2\pi)$.

- $\sin^{-1} (-3\sqrt{2})$
- $\cos^{-1} (-2\sqrt{2})$
- $\tan^{-1} \sqrt{3}$

Solution:

- Recall that $-3\sqrt{2}$ is from the 30−60−90 triangle. The reference angle for sin and $3\sqrt{2}$ would be 60°. Because this is sine and it is negative, it must be in the third or fourth quadrant. The answer is either $4\pi/3$ or $5\pi/3$.
- $-2\sqrt{2}$ is from an isosceles right triangle. The reference angle is then 45°. Because this is cosine and negative, the angle must be in either the second or third quadrant. The answer is either $3\pi/4$ or $5\pi/4$.
- $\sqrt{3}$ is also from a 30−60−90 triangle. Tangent is $\sqrt{3}$ for the reference angle 60°. Tangent is positive in the first and third quadrants, so the answer would be $\pi/3$ or $4\pi/3$.

Notice how each one of these examples yields two answers. This poses a problem when finding a singular inverse for each of the trig functions. Therefore, we need to restrict the domain in which the inverses can be found.

Example: The value of 6 $\sin^{-1} 1$.

Solution: Let, $A = \sin^{-1} 1$, then $\sin A = 1$

Since,

$$\text{Sin}\frac{\pi}{2} = 1,$$

$$6\sin^{-1}1 = 6 \times \frac{\pi}{2}$$

$$6\sin^{-1}1 = 3\pi$$

Example: Find the value of $\tan^{-1}(1.1106)$.

Solution: Let, $A = \tan^{-1}(1.1106)$

Then,

$$\tan A = 1.1106$$

$$A = 48°$$

$$\tan 48 = 1.1106$$

(Use calculator in degree mode)

$$\tan^{-1}1.1106 = 48°$$

Example: $\sin^{-1}(\cos \pi/3) = ?$

Solution:

$$\sin^{-1}\left(\cos\frac{\pi}{3}\right) = \sin^{-1}\frac{1}{2} \quad \left[\text{substitute } \cos\left(\frac{\pi}{3}\right) = \frac{1}{2}\right]$$

$$= \frac{\pi}{6} \quad \left[\text{substitute } \sin^{-1}\left(\frac{1}{2}\right) = \frac{\pi}{6}\right]$$

Example: Find the value of $\sin\left(\frac{\pi}{4} + \text{Cos}^{-1}\left(\frac{\sqrt{2}}{2}\right)\right)$.

Solution: Let, $y = \sin\left(\frac{\pi}{4} + \text{Cos}^{-1}\left(\frac{\sqrt{2}}{2}\right)\right)$ and $A = \text{Cos}^{-1}\left(\frac{\sqrt{2}}{2}\right)$

Then,

$$A = \text{Cos}\frac{\sqrt{2}}{2}$$

Multiplying the numerator as well as denominator by $\sqrt{2}$ we get:

$$\text{Cos}\,A = \frac{1}{\sqrt{2}}$$

$$A = \frac{\pi}{4}$$

Therefore,

$$y = \sin\left(\frac{\pi}{4} + \frac{\pi}{4}\right)$$

$$y = \sin\left(\frac{\pi}{2}\right)$$

Hence, y=1.

PROPERTIES OF INVERSE TRIGONOMETRIC FUNCTIONS

Here are the properties of the inverse trigonometric functions with proof.

Property:

- $\sin^{-1}(1/x) = \text{cosec}^{-1}x,\ x \geq 1 \text{ or } x \leq -1$
- $\cos^{-1}(1/x) = \sec^{-1}x,\ x \geq 1 \text{ or } x \leq -1$
- $\tan^{-1}(1/x) = \cot^{-1}x,\ x > 0$

Proof: $\sin^{-1}(1/x) = \text{cosec}^{-1}x,\ x \geq 1 \text{ or } x \leq -1$

Let, $\sin^{-1}x = y$

i.e. $x = \text{cosec}\,y$

$$\frac{1}{x} = \sin y$$

$$\sin^{-1}\left(\frac{1}{x}\right) = y$$

$$\sin^{-1}\left(\frac{1}{x}\right) = \text{cosec}^{-1}x$$

$$\sin^{-1}\left(\frac{1}{x}\right) = \text{cosec}^{-1}x$$

Hence, $\sin^{-1}\left(\frac{1}{x}\right) = \text{cosec}^{-1}x$,

where, $x \geq 1$ or $x \leq -1$.

Property:

- $\sin^{-1}(-x) = -\sin^{-1}(x), \quad x \in [-1,1]$
- $\tan^{-1}(-x) = -\tan^{-1}(x), \ x \in R$
- $\text{cosec}^{-1}(-x) = -\text{cosec}^{-1}(x), \ |x| \geq 1$

Proof: $\sin^{-1}(-x) = -\sin^{-1}(x), \quad x \in [-1,1]$

Let, $\sin^{-1}(-x) = y$

Then, $-x = \sin y$

$$x = -\sin y$$
$$x = \sin(-y)$$
$$\sin^{-1} = \sin^{-1}(\sin(-y))$$
$$\sin^{-1}x = y$$
$$\sin^{-1}x = -\sin^{-1}(-x)$$

Hence, $\sin^{-1}(-x) = -\sin^{-1}x \in [-1,1]$.

Property:

- $\cos^{-1}(-x) = \pi - \cos^{-1}x, \ x \in [-1,1]$
- $\sec^{-1}(-x) = \pi - \sec^{-1}x, \ |x| \geq 1$
- $\cot^{-1}(-x) = \pi - \cot^{-1}x, \ x \in R$

Proof: $\cos^{-1}(-x) = \pi - \cos^{-1}x, \ x \in [-1,1]$

Let, $\cos^{-1}(-x) = y$

$$\cos y = -x \quad x = -\cos y$$
$$x = \cos(\pi - y)$$

Since, $\cos\pi - q = -\cos q$

$$\cos^{-1}x = \pi - y$$
$$\cos^{-1}x = \pi - \cos^{-1} - x$$

Hence, $\cos^{-1} - x = \pi - \cos^{-1}x$.

Property:

- $\sin^{-1}x + \cos^{-1}x = \pi/2,\ x \in [-1,1]$
- $\tan^{-1}x + \cot^{-1}x = \pi/2,\ x \in R$
- $\mathrm{cosec}^{-1}x + \sec^{-1}x = \pi/2,\ |x| \geq 1$

Proof: $\sin^{-1}x + \cos^{-1}x = \pi/2,\ x \in [-1,1]$

Let, $\sin^{-1}x = y$ or $x = \sin y = \cos\left(\frac{\pi}{2} - y\right)$

$$\cos^{-1}x = \cos^{-1}\left(\cos\left(\frac{\pi}{2} - y\right)\right)$$
$$\cos^{-1}x = \frac{\pi}{2} - y$$
$$\cos^{-1}x = \frac{\pi}{2} - \sin^{-1}x$$
$$\sin^{-1} + \cos^{-1}x = \frac{\pi}{2}$$

Hence, $\sin^{-1}x + \cos^{-1}x = \pi/2,\ x \in [-1,1]$.

Property:

- $\tan^{-1}x + \tan^{-1}y = \tan^{-1}((x+y)/(1-xy)),\ xy < 1$
- $\tan^{-1}x - \tan^{-1}y = \tan^{-1}((x-y)/(1+xy)),\ xy > -1$

Proof: $\tan^{-1}x + \tan^{-1}y = \tan^{-1}((x+y)/(1-xy)),\ xy < 1$

Let, $\tan^{-1}x = A$

And $\tan^{-1}y = B$

Then, $\tan A = x$

$\tan B = y$

Now, $\tan(A+B) = (\tan A + \tan B)/(1 - \tan A \tan B)$

$$\tan(A+B)=\frac{x+y}{1-xy}$$

$$\tan^{-1}\left(\frac{x+y}{1-xy}\right)=A+B$$

Hence, $\tan^{-1}\left(\frac{x+y}{1-xy}\right)=\tan^{-1}x+\tan^{-1}y.$

Property:

- $2\tan^{-1}x = \sin^{-1}(2x/(1+x^2)),\ |x| \le 1$
- $2\tan^{-1}x = \cos^{-1}((1-x^2)/(1+x^2)),\ x \ge 0$
- $2\tan^{-1}x = \tan^{-1}(2x/(1-x^2)),\ -1 < x < 1$

Proof: $2\tan^{-1}x = \sin^{-1}(2x/(1+x^2)),\ |x| \le 1$

Let, $\tan^{-1}x=y$ and $x=\tan y$

Consider RHS: $\sin^{-1}\left(\frac{2x}{1+x^2}\right)$

$$=\sin^{-1}\left(\frac{2\tan y}{1+\tan^2 y}\right)$$

$$=\sin^{-1}(\sin 2y)$$

Since, $\sin 2\theta=2\tan\theta/(1+\tan^2\theta)$,

$$=2y$$

$$=2\tan^{-1}x, \text{ which is our LHS.}$$

Hence, $2\tan^{-1}x=\sin^{-1}(2x/(1+x^2)),\ |x| \le 1.$

Example: Prove that "$\sin^{-1}(-x) = -\sin^{-1}(x),\ x \in [-1,1]$"

Solution: Let, $\sin^{-1}(-x)=y$

Then, $-x=\sin y$

$$x=-\sin y$$

$$x=\sin(-y)$$

$$\sin^{-1}x=\arcsin(\sin(-y))$$

$$\sin^{-1}x=y$$

$$\sin^{-1}x = -\sin^{-1}(-x)$$

Hence, $\sin^{-1}(-x) = -\sin^{-1}x,\ x \in [-1,1]$.

ARCSINE

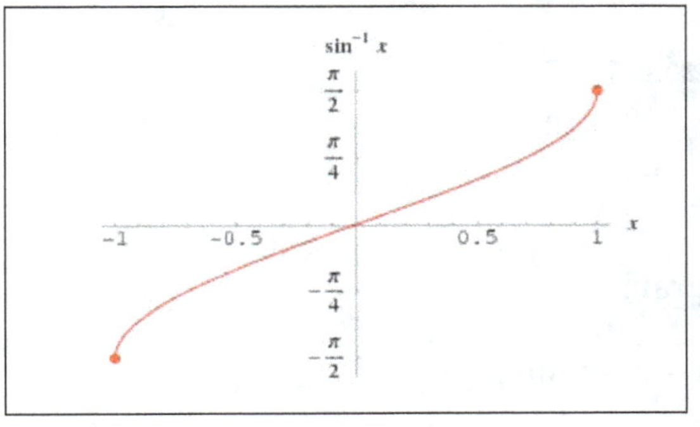

The inverse sine is the multivalued function $\sin^{-1} z$, also denoted arcsin z, that is the inverse function of the sine. The variants Arcsin z and $\mathrm{Sin}^{-1} z$ are sometimes used to refer to explicit principal values of the inverse sine, although this distinction is not always made. Worse yet, the notation arcsin z is sometimes used for the principal value, with Arcsin z being used for the multivalued function. Note that in the notation $\sin^{-1} z$ (commonly used in North America and in pocket calculators worldwide), sin z is the sine and the superscript -1 denotes the inverse function, not the multiplicative inverse.

The principal value of the inverse sine is implemented as Arcsin z. In the GNUC library, it is implemented as asin(double x).

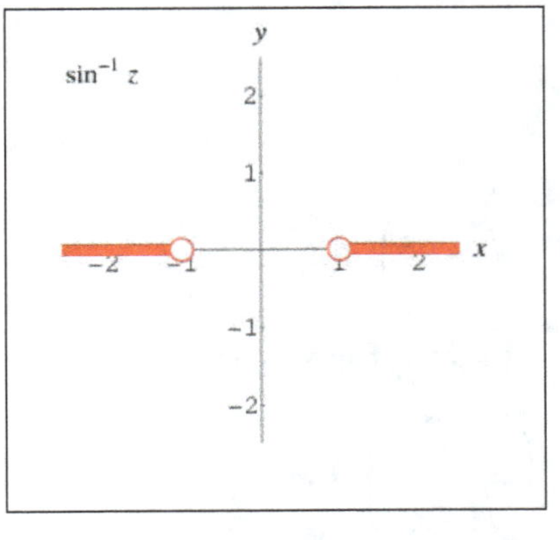

The inverse sine is a multivalued function and hence requires a branch cut in the complex plane. This follows from the definition of $\sin^{-1} z$ as:

$$\sin^{-1}z = -i\ \ln\left(iz+\sqrt{1-z^2}\right).$$

Special values include:

$$\sin^{-1}(-1) = -\frac{1}{2}\pi$$

$$\sin^{-1}0 = 0$$

$$\sin^{-1}1 = \frac{1}{2}\pi.$$

The derivative of $\sin^{-1} z$ is:

$$\frac{d}{dz}\sin^{-1} z = \frac{1}{\sqrt{1-z^2}},$$

and its indefinite integral is:

$$\int \sin^{-1} z\, dz = \sqrt{1-z^2} + z\sin^{-1} z + C.$$

The inverse sine satisfies:

$$\sin^{-1} z = \csc^{-1}\left(\frac{1}{z}\right).$$

For $z \neq 0$,

$$\begin{aligned}\sin^{-1} z &= \sin^{-1}(-z) \\ &= \cos^{-1}(-z) - \frac{1}{2}\pi \\ &= \frac{1}{2}\pi - \cos^{-1} z\end{aligned}$$

For all complex z,

$$\begin{aligned}\sin^{-1} x &= \begin{cases} -\frac{1}{2}\pi + \sin^{-1}\left(\sqrt{1-x^2}\right) & \text{for } x < 0 \\ \frac{1}{2}\pi - \sin^{-1}\left(\sqrt{1-x^2}\right) & \text{for } x > 0 \end{cases} \\ &= \begin{cases} -\frac{1}{2}\pi - \cot^{-1}\left(\frac{x}{\sqrt{1-x^2}}\right) & \text{for } x < 0 \\ \frac{1}{2}\pi - \cot^{-1}\left(\frac{x}{\sqrt{1-x^2}}\right) & \text{for } x > 0 \end{cases}\end{aligned}$$

$$= \begin{cases} -\cos^{-1}\left(\sqrt{1-x^2}\right) & \text{for } -1 < x < 0 \\ \cos^{-1}\left(\sqrt{1-x^2}\right) & \text{for } 0 < x < 1 \end{cases}$$

$$= \begin{cases} -\sec^{-1}\left(\frac{1}{\sqrt{1-x^2}}\right) & \text{for } -1 < x < 0 \\ \sec^{-1}\left(\frac{1}{\sqrt{1-x^2}}\right) & \text{for } 0 < x < 1, \end{cases}$$

and,

$$\sin^{-1} x = \tan^{-1}\left(\frac{x}{\sqrt{1-x^2}}\right)$$

$$= \cot^{-1}\left(\frac{\sqrt{1-x^2}}{x}\right),$$

for $-1 < x < 1$, where equality at points where the denominators are 0 is understood to mean in the limit as $x \to \pm 1$ or $x \to 0$, respectively.

The Maclaurin series for the inverse sine with $-1 \le x \le 1$ is given by:

$$\sin^{-1} x = \sum_{n=0}^{\infty} \frac{\left(\frac{1}{2}\right)_n}{(2n+1)n!} x^{2n+1}$$

$$= x + \frac{1}{6}x^3 + \frac{3}{40}x^5 + \frac{5}{112}x^7 + \frac{35}{1152}x^9 + \ldots$$

Where, x_n is a Pochhammer symbol.

The inverse sine can be given by the sum:

$$\left(\sin^{-1} x\right)^2 = \frac{1}{2}\sum_{n=1}^{\infty} \frac{(2x)^{2n}}{n^2 \binom{2^n}{n}}$$

Where,

$\binom{2n}{n}$ is a binomial coefficient. Similarly,

$$\left[\sin^{-1}\left(\frac{1}{2}x\right)\right]^4 = \frac{3}{2}\sum_{k=1}^{\infty}\left[\sum_{m=1}^{k-1}\frac{1}{m^2}\right]\frac{x^{2k}}{k^2\binom{2k}{k}}$$

$$\left[\sin^{-1}\left(\frac{1}{2}x\right)\right]^6 = \frac{45}{4}\sum_{k=1}^{\infty}\left[\sum_{m=1}^{k-1}\frac{1}{m^2}\sum_{n=1}^{m-1}\frac{1}{n^2}\right]\frac{x^{2k}}{k^2\binom{2k}{k}}$$

$$\left[\sin^{-1}\left(\frac{1}{2}x\right)\right]^8 = \frac{315}{2}\sum_{k=1}^{\infty}\left[\sum_{m=1}^{k-1}\frac{1}{m^2}\sum_{n=1}^{m-1}\frac{1}{n^2}\sum_{p=1}^{n-1}\frac{1}{p^2}\right]\frac{x^{2k}}{k^2\binom{2k}{k}}$$

Ramanujan gave the cases $(\sin^{-1}x)^n$ for n=1, 2, 3, and 4, and the general cases are given in terms of multiple sums by Bailey et al. and Borwein and Chamberland.

The inverse sine has continued fraction:

$$\sin^{-1}z = \cfrac{z\sqrt{1-z^2}}{1-\cfrac{1.2z^2}{3-\cfrac{1.2z^2}{5-\cfrac{3.4z^2}{7-\cfrac{3.4z^2}{9-\cfrac{5.6z^2}{11-\ldots}}}}}}$$

Arcsine of Zero

$$\sin 0 = \sin 0^\circ = 0$$

The arcsine of 0 is equal to the inverse sine function of 0, which is equal to 0 radians or 0 degrees:

$$\arcsin 0 = \sin^{-1} 0 = 0 \text{ rad} = 0^\circ$$

Arcsine of 1

$$\sin \pi/2 = \sin 90^\circ = 1$$

The arcsine of 1 is equal to the inverse sine function of 1, which is equal to $\pi/2$ radians or 90 degrees:

$$\arcsin 1 = \sin^{-1} 1 = \pi/2 \text{ rad} = 90^\circ$$

Arcsin of Infinity

The arcsine is the inverse sine function.

Since x can be in the range of [-1, 1], arcsin(x) is undefined outside the range of [-1, 1].

So the limit of arcsine of x when x is approaching infinity is undefined:

$$\lim_{x\to\infty} \arcsin(x) = \text{undefined}$$

ARCCOSINE

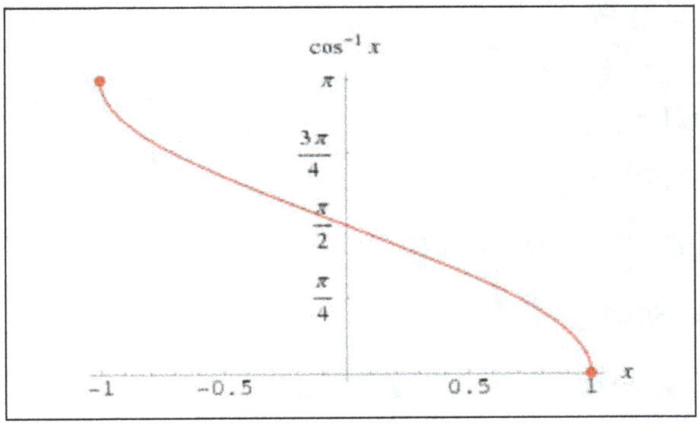

The inverse cosine is the multivalued function $\cos^{-1} z$, also denoted arcos z, that is the inverse function of the cosine. The variants Arccos z and Cos^{-1} z are sometimes used to refer to explicit principal values of the inverse cosine, although this distinction is not always made. Worse yet, the notation arccosz is sometimes used for the principal value, with Arccos z being used for the multivalued function. Note that thenotation $\cos^{-1} z$ (commonly used in North Americaand in pocket calculators worldwide), cos z is the cosine and the superscript -1 denotes the inverse function, not the multiplicative inverse.

The principal value of the inverse cosine is implemented as arccos[z].

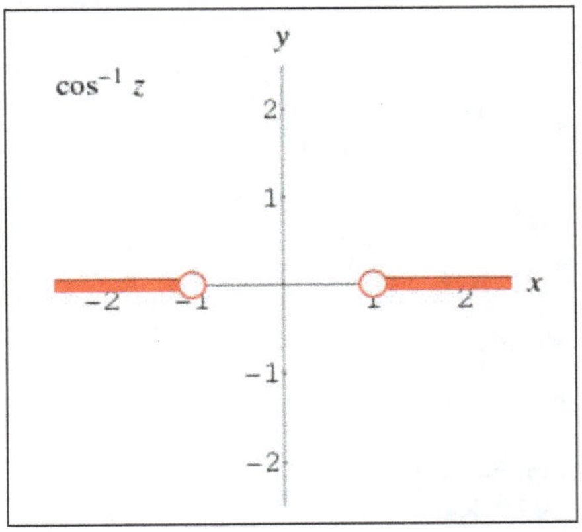

The inverse cosine is a multivalued function and hence requires a branch cut in the complex plane, which the Wolfram Language's convention places at the line segments (-∞, -1) and (1, ∞). This follows from the definition of $\cos^{-1} z$ as:

$$\cos^{-1} z = \frac{1}{2}\pi + i \ln\left(iz + \sqrt{1-z^2}\right).$$

Special values include:

$$\cos^{-1}(-1) = \pi$$

$$\cos^{-1} 0 = \frac{1}{2}\pi$$

$$\cos^{-1} 1 = 0.$$

The derivative of $\cos^{-1} z$ is given by:

$$\frac{d}{dz}\cos^{-1} z = -\frac{1}{\sqrt{1-z^2}}.$$

And its indefinite integral is:

$$\int \cos^{-1} z\, dz = z\cos^{-1} z - \sqrt{1-z^2} + C.$$

The inverse cosine satisfies:

$$\cos^{-1} z = \pi - \cos^{-1}(-z).$$

For all complex z, and

$$\cos^{-1} x = \begin{cases} \frac{1}{2}\pi + \cos^{-1}\left(\sqrt{1-x^2}\right) & \text{for } x \le 0 \\ \frac{1}{2}\pi - \cos^{-1}\left(\sqrt{1-x^2}\right) & \text{for } x \ge 0. \end{cases}$$

The inverse cosine is given in terms of other inverse trigonometric functions by:

$$\cos^{-1} z = \frac{1}{2}\pi + \sin^{-1}(-z)$$

$$= \frac{1}{2}\pi - \sin^{-1} z.$$

For all complex z,

$$\cos^{-1} z = \sec^{-1}\left(\frac{1}{z}\right).$$

For $z \ne 0$,

$$\cos^{-1} x = \frac{1}{2}\pi - \tan^{-1}\left(\frac{x}{\sqrt{1\ \ x}}\right).$$

For $-1 \le x \le 1$, and

$$\cos^{-1} x = \cot^{-1}\left(\frac{x}{\sqrt{1-x^2}}\right)$$
$$= \csc^{-1}\left(\frac{1}{\sqrt{1-x^2}}\right)$$
$$= \sin^{-1}\left(\sqrt{1-x^2}\right)$$
$$= \tan^{-1}\left(\frac{\sqrt{1-x^2}}{x}\right).$$

For $x \ge 0$, where in the last equation, equality at zero is understood to mean in the limit as $x \to 0^+$.

The Maclaurin series for the inverse cosine with $-1 \le x \le 1$ is:

$$\cos^{-1} x = \frac{1}{2}\pi - \sum_{n=1}^{\infty} \frac{\binom{\frac{1}{2}}{2}_{n-1}}{(n-1)!(2n-1)} x^{2n-1} = \frac{1}{2}\pi - x - \frac{1}{6}x^3 - \frac{3}{40}x^5 - \frac{5}{112}x^7 - \frac{35}{1152}x^9 - \ldots$$

ARCTANGENT

The arc tangent function, denoted arctan or $\tan^{-1}$, is a function defined as follows: for $x \in \mathbb{R}$, arctan x is the unique number y in the open interval $(-\pi/2, \pi/2)$ such that tan y = x.

Equivalently, the arc tangent function is the inverse function to the restriction of the tangent function to the interval $(-\pi/2, \pi/2)$.

Key Data

Item	Value
Default domain	All real numbers, i.e., all of $\mathbb{R}$
Range	The open interval $(-\pi/2, \pi/2)$, i.e., the set $\{y \mid -\pi/2 < y < \pi/2\}$ no absolute minimum or absolute maximum, because the extremes are attained asymptotically at infinity.
Local minimum values and points of attainment	No local minimum values.
Local maximum values and points of attainment	No local maximum values.
Points of inflection (both coordinates)	(0, 0) (the origin) only.

Horizontal asymptotes	The line $y = \pi/2$ as $x \to \infty$ The line $y = -\pi/2$ as $x \to -\infty$
Derivative	$x \mapsto \frac{1}{1+x^2}$
Second derivative	$x \mapsto \frac{-2x}{(1+x^2)^2}$
Higher derivatives	The nth derivative is a rational function of x where the denominator degree is n+1 more than the numerator degree.
Antiderivative	$x \arctan x - \frac{1}{2}\ln(1+x^2) + C$. We use integration by parts.
Higher antiderivatives	The function can be anti-differentiated any number of times in terms of elementary functions.

Graph

Here is a graph of arctan from a zoomed out position, where the horizontal asymptotes are clear.

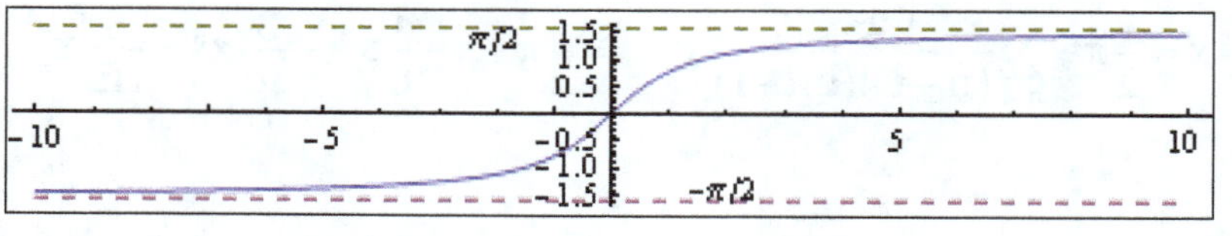

Here is a more close-up version, with the tangent line through the origin (the line y = x) drawn, indicating that the origin is a point of inflection for the graph:

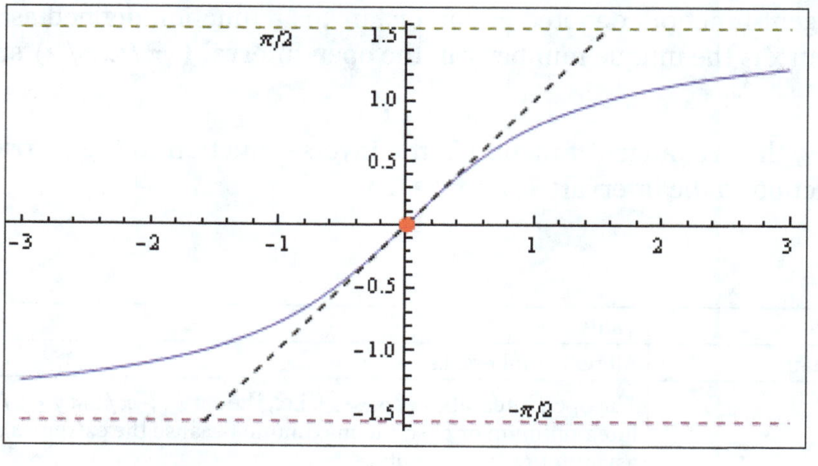

Differentiation

First Derivative

We use the inverse function theorem, and the fact that the derivative of tan is $\sec^2$.

By the inverse function theorem, we have:

$$\arctan' x = \frac{1}{\tan'(\arctan x)} = \frac{1}{\sec^2(\arctan x)}$$

If $\arctan x = \theta$, then, $x = \tan\theta$ and we get:

$$\sec^2\theta = 1 + \tan^2\theta = 1 + x^2$$

Plugging this into the above, we get:

$$\arctan' x = \frac{1}{1+x^2}$$

Second Derivative

The second derivative is given as:

$$\arctan'' x = \frac{d}{dx}\arctan' x = \frac{d}{dx}\left(\frac{1}{1+x^2}\right) = \frac{d}{dx}[(1+x^2)^{-1}]:$$

$$= (2x)(-1)(1+x^2)^{-2} = \frac{-2x}{(1+x^2)^2}$$

Higher Derivatives

For higher derivatives, we use the quotient rule for differentiation, combined with the chain rule for differentiation to deal with powers of $1+x^2$.

Integration

First Antiderivative

We can integrate this using the inverse function integration method, and obtain:

$$x\arctan x - \int \tan u \; du \quad u = \arctan x$$

This becomes:

$$x\arctan x + \ln|\cos u| + C$$

We have $\ln|\cos u| = (1/2)\ln(\cos^2 u) = -1/2\ln(\sec^2 u) = (-1/2)\ln(1+x^2)$, and we get:

$$x\arctan x - \frac{1}{2}\ln(1+x^2) + C$$

More explicitly, we can do the integration using integration by parts taking arctan x as the part to differentiate and 1 as the part to integrate:

$$x \arctan x - \int \frac{1}{1+x^2} \cdot x dx$$

For the second integration, we integrate using the g'/g formulation to get:

$$(1/2)\ln(1+x^2)$$

Higher Antiderivatives

The function can be antidifferentiated any number of times using integration by parts. The reason for this is that the derivative of the function is a rational function, and rational functions can be repeatedly integrated within elementarily expressible functions.

All the antiderivatives can be expressed in the form:

$$p_0(x) + p_1(x) \arctan x + p_2(x) \ln(1+x^2)$$

Where, p_0, p_1, p_2 are polynomial. Note that p_0 is ambiguous up to addition of polynomials of degree n - 1 if we are integrating n times.

The function can be antidifferentiated any number of times using integration by parts.

Power Series and Taylor Series

Computation of Power Series

The power series for the function about 0 can be obtained as follows:

We know that for the function $1/(1+x^2)$, we have the power series:

$$\frac{1}{1+x^2} = 1 - x^2 + x^4 - x^6 + \ldots, \qquad -1 < x < 1$$

Integrating with a definite integral, we get:

$$\int_0^x \frac{1}{1+t^2} dt = x - \frac{x^3}{3} + \frac{x^5}{5} - \frac{x^7}{7} + \ldots, -1 < x < 1$$

The left side is $[\arctan t]_0^x = \arctan x - \arctan 0 = \arctan x$, so we get:

$$\arctan x = x - \frac{x^3}{3} + \frac{x^5}{5} - \frac{x^7}{7} + \ldots = \sum_{k=0}^{\infty} \frac{(-1)^k x^{2k+1}}{2k+1}, \qquad -1 < x < 1$$

By the alternating series theorem, we note that the power series on the right converges for x = -1 and for x = 1, so by Abel's theorem, we conclude that it converges to arctan of the respective inputs. We thus get:

$$\arctan x = x - \frac{x^3}{3} + \frac{x^5}{5} - \frac{x^7}{7} + \ldots = \sum_{k=0}^{\infty} \frac{(-1)^k x^{2k+1}}{2k+1}, \qquad -1 \le x \le 1$$

ARCCOSECANT

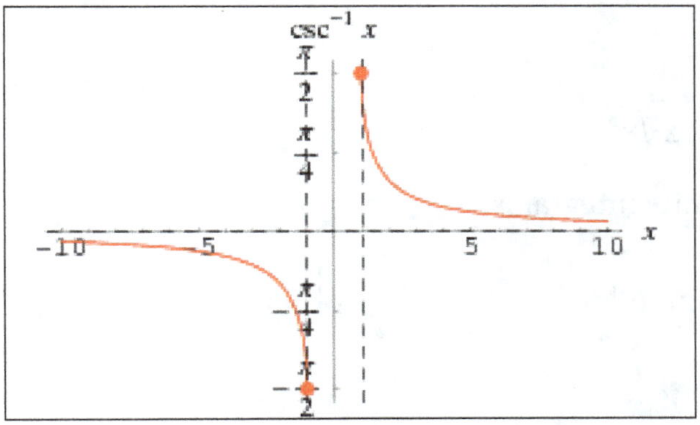

The inverse cosecant is the multivalued function $\csc^{-1} z$, also denoted arccsc z, that is the inverse function of the cosecant. The variants Arccsc z and $\mathrm{Csc}^{-1} z$ are sometimes used to refer to explicit principal values of the inverse cosecant, although this distinction is not always made. Worse yet, the notation arccsc z is sometimes used for the principal value, with Arccsc z being used for the multivalued function. Note that in the notation $\csc^{-1} z$ (commonly used in North America and in pocket calculators worldwide), csc z is the cosecant and the superscript -1 denotes an inverse function, not the multiplicative inverse.

The principal value of the inverse cosecant is implemented as arccsc x in the Wolfram Language.

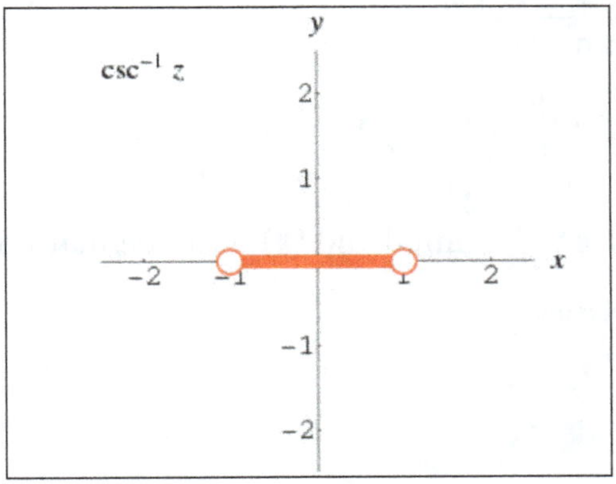

The inverse cosecant is a multivalued function and hence requires a branch cut in the complex plane. This follows from the definition of $\csc^{-1} z$ as:

$$\csc^{-1}z = -i\ln\left(\sqrt{1-\frac{1}{z^2}}+\frac{i}{z}\right).$$

The derivative of $\csc^{-1} z$ is given by:

$$\frac{d}{dz}\csc^{-1}z = -\frac{1}{z^2\sqrt{1-\frac{1}{z^2}}},$$

which simplifies to:

$$\frac{d}{dx}\csc^{-1}x = -\frac{1}{x\sqrt{x^2-1}}.$$

For x > o. Its indefinite integral is:

$$\int \csc^{-1}z\,dz = z\csc^{-1}z + \ln\left[z\left(1+\sqrt{\frac{z^2-1}{z^2}}\right)\right]+C,$$

which simplifies to:

$$\int \csc^{-1}x\,dx = x\csc^{-1}x + \ln\left(x+\sqrt{x^2-1}\right).$$

For x > o.

The inverse cosecant has Taylor series about infinity of

$$\begin{aligned}\csc^{-1}x &= -\sum_{n=1}^{\infty}\frac{i^{n+1}P_{n-1}(o)}{n}x^{-n}\\ &= \frac{\left(\frac{1}{2}\right)_{n-1}}{(n-1)!(2n-1)}x^{1-2n}\\ &= x^{-1}+\frac{1}{6}x^{-3}+\frac{3}{40}x^{-5}+\frac{5}{112}x^{-7}+\ldots\end{aligned}$$

Where, $P_n(x)$ is a Legendre polynomial and $(x)_n$ is a Pochhammer symbol.

The inverse cosecant satisfies:

$$\csc^{-1}z = \sin^{-1}\left(\frac{1}{z}\right)$$

For $z \neq 0$,

$$\csc^{-1} z = \frac{1}{2}\pi - \sec^{-1} z = -\frac{1}{2}\pi + \sec^{-1}(-z)$$

For all complex z, and

$$\csc^{-1} x = \begin{cases} \sec^{-1}\left(\frac{x}{\sqrt{x^2-1}}\right) - \pi & \text{for } x<-1 \\ \sec^{-1}\left(\frac{x}{\sqrt{x^2-1}}\right) & \text{for } x>1 \end{cases}$$

$$= \begin{cases} -\cos^{-1}\left(\frac{\sqrt{x^2-1}}{x}\right) & \text{for } x<-1 \\ \cos^{-1}\left(\frac{\sqrt{x^2-1}}{x}\right) & \text{for } x>1 \end{cases} = \begin{cases} -\cot^{-1}\left(\sqrt{x^2-1}\right) & \text{for } x<-1 \\ \cot^{-1}\left(\sqrt{x^2-1}\right) & \text{for } x>1 \end{cases}$$

ARCSECANT

From trigonometry, we know that the secant function is defined by:

$$\sec(x) = \frac{1}{\cos(x)}$$

Since cos(x)= 0 at x $\pm\pi/2$, $\pm 3\pi/2$, $\pm 5\pi/2$. etc. (all odd multiples of $\pi/2$), the secant function is not defined at these values of x (since division by zero is undefined). Also, since the range of the cosine function is [-1, 1], the range of the secant function is [-∞, -1] ∪ [1, ∞]. A portion of the graph of the secant function. The graph has vertical asymptotes at all odd multiples of $\pi/2$.

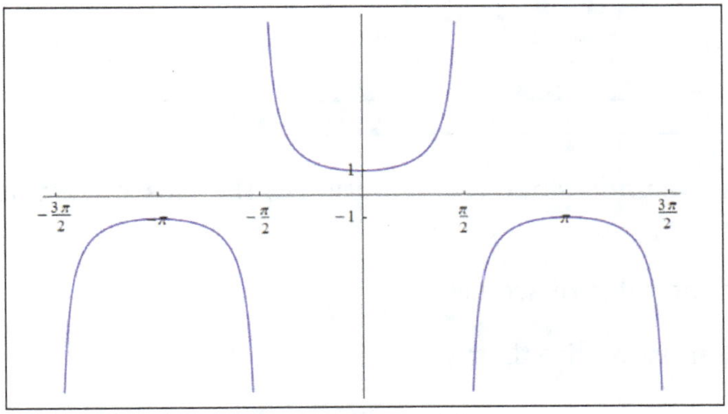

Graph of y = sec(x).

If we restrict the domain of the secant function to [o, π /2) ∪ (π/2, π] (as shown in the graph), then we have a one–to–one function which thus has an inverse.

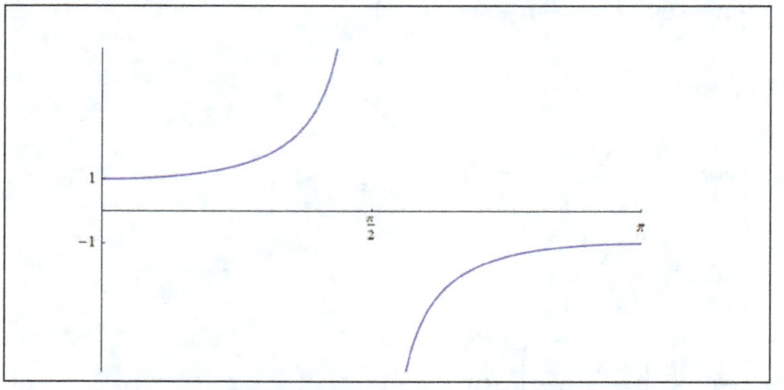

y = sec(x) with domain [o, π/2) ∪ (π/2, π).

The inverse secant function (denoted by arcsec) is defined using the restricted domain on the secant function described above. Thus the inverse secant function has domain [-∞, -1] ∪ [1, ∞] and range [o, π /2) ∪ (π/2, π]. For each x in the set [-∞, -1] ∪ [1, ∞] we define:

y = arcsec(x)

To be the unique value of y in the set [o, π /2) ∪ (π/2, π] for which:

sec(y) = x.

A graph of the inverse secant function is shown.

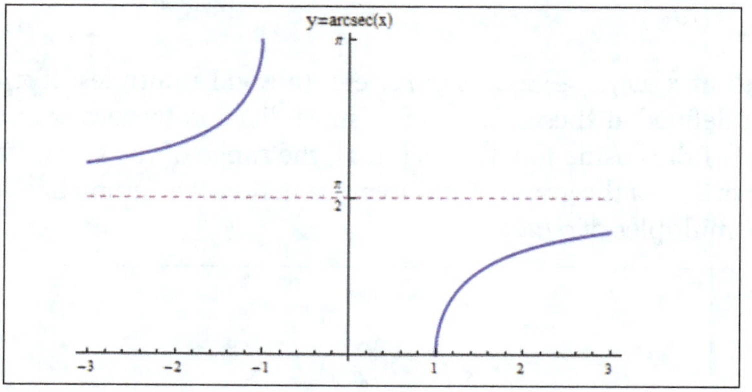

To make sure that we understand the definition of the arcsec function, let us consider a couple of examples.

Example: What is the value of arcsec(1)?

To answer this we must realize that:

y = arcsec(1)

Means that:

$$\sec(y) = 1$$

And that y is a number in $[0, \pi/2) \cup (\pi/2, \pi]$. Based on our knowledge of trigonometric functions we know that $\cos(0) = 1$ and hence:

$$\sec(0) = \frac{1}{\cos(0)} = \frac{1}{1} = 1.$$

Furthermore we see that 0 is a number in the set $[0, \pi/2) \cup (\pi/2, \pi]$. Therefore arcsec(1) = 0.

Why isn't it true that arcsec(1) = 2π? After all it is true that:

$$\sec(2\pi) = \frac{1}{\cos(2\pi)} = \frac{1}{1} = 1.$$

However note that the number 2π is not in the set $[0, \pi/2) \cup (\pi/2, \pi]$. Thus arcsec $1 \neq 2\pi$ even though $\sec(2\pi) = 1$.

Example: What is the value of $\operatorname{arc}\sec\left(-\sqrt{2}\right)$?

To answer this we must realize that:

$$y = \operatorname{arc}\sec\left(-\sqrt{2}\right),$$

means that:

$$\sec(y) = \left(-\sqrt{2}\right),$$

and that y is a number in the set $[0, \pi/2) \cup (\pi/2, \pi]$. Based on our knowledge of trigonometric functions we know that $\cos(3\pi/4) = -\sqrt{2}/2$ and hence:

$$\cos(3\pi/4) = \frac{1}{\cos\left(\frac{3\pi}{4}\right)} - \frac{1}{-\frac{\sqrt{2}}{2}} = -\sqrt{2}$$

Furthermore we see that $3\pi/4$ is a number in the set $[0, \pi/2) \cup (\pi/2, \pi]$. Therefore

$$\operatorname{arcsec}(-\sqrt{2}) = 3\pi/4.$$

Why isn't it true that $\operatorname{arcsec}(-\sqrt{2}) = 3\pi/4$? After all it is true that:

$$\sec\left(-\frac{3\pi}{4}\right) = \frac{1}{\cos\left(-\frac{3\pi}{4}\right)} = \frac{1}{-\frac{\sqrt{2}}{2}} = -\sqrt{2}$$

However note that the number $-3\pi/4$ is not in the set $[0, \pi/2) \cup (\pi/2, \pi]$. Thus arcsec $(-\sqrt{2}) \neq -3\pi/4$ even though sec $(-3\pi/4) = (-\sqrt{2})$. In fact, notice that the value of arcsec (x) is never negative for any value of x.

The Derivative of Arcsec

We now compute the derivative of y = arcsec (x). Rewriting y = arcsec (x) as sec (y) = x and using implicit differentiation gives:

$$\frac{d}{dx}(\sec(y)) = \frac{d}{dx}(x),$$

or,

$$\sec(y)\tan(y)\frac{dy}{dx} = 1,$$

or,

$$\frac{dy}{dx} = \frac{1}{\sec(y)\tan(y)}.$$

Using the trigonometric identity:

$$\sec^2(y) = 1 + \tan^2(y),$$

we obtain:

$$\tan(y) = \pm\sqrt{\sec^2(y) - 1}.$$

If x is in the interval $(1, \infty)$, then y is in the interval $(0, \pi/2)$ meaning that tan (y) > 0 and we choose the "+" sign on the above square root. This gives us:

$$\frac{dy}{dx} = \frac{1}{-x\sqrt{x^2 - 1}}.$$

At x = 1 and x = −1, the arcsec function is not differentiable. Its graph has a vertical (infinite slope) tangent line at these points.

In conclusion, the derivative of the inverse secant function is:

$$\frac{d}{dx}(\operatorname{arcsec}(x)) = \begin{cases} \dfrac{1}{-x\sqrt{x^2 - 1}} & \text{if } -\infty < x < -1 \\ \dfrac{1}{x\sqrt{x^2 - 1}} & \text{if } 1 < x < \infty \end{cases}.$$

Integrals involving Arcsec

From the above differentiation formula, we can see that the value of:

$$\int \frac{1}{x\sqrt{x^2-1}}dx,$$

depends on what interval we are working with. If we are working on some interval that is a subset of (1, ∞), then:

$$\int \frac{1}{x\sqrt{x^2-1}}dx = \operatorname{arcsec}(x)+C.$$

However if we are working on some interval that is a subset of (−∞,−1), then:

$$\int \frac{1}{x\sqrt{x^2-1}}dx = -\operatorname{arcsec}(x)+C,$$

because on such an interval we have:

$$\frac{d}{dx}(-\operatorname{arcsec}(x)) = -\frac{d}{dx}(\operatorname{arcsec}(x)) = -\left(\frac{1}{-x\sqrt{x^2-1}}\right) = \frac{1}{x\sqrt{x^2-1}}.$$

For example, the interval $\left[\sqrt{2}, 2\right]$ is a subset of $[1,\infty]$ and thus:

$$\int_{\sqrt{2}}^{2} \frac{1}{x\sqrt{x^2-1}}dx = \operatorname{arcsec}(x)\Big|_{\sqrt{2}}^{2} = \operatorname{arcsec}(2) - \operatorname{arcsec}\left(\sqrt{2}\right) = \frac{\pi}{12},$$

however, the interval $\left[-2, -\sqrt{2}\right]$ is a subset of (−∞, −1) and hence:

$$\int_{-2}^{-\sqrt{2}} \frac{1}{x\sqrt{x^2-1}}dx = -\operatorname{arcsec}(x)\Big|_{2}^{-\sqrt{2}} = -\operatorname{arcsec}\left(-\sqrt{2}\right) - \left(-\operatorname{arcsec}(-2)\right) = -\frac{\pi}{12}.$$

ARCCOTANGENT

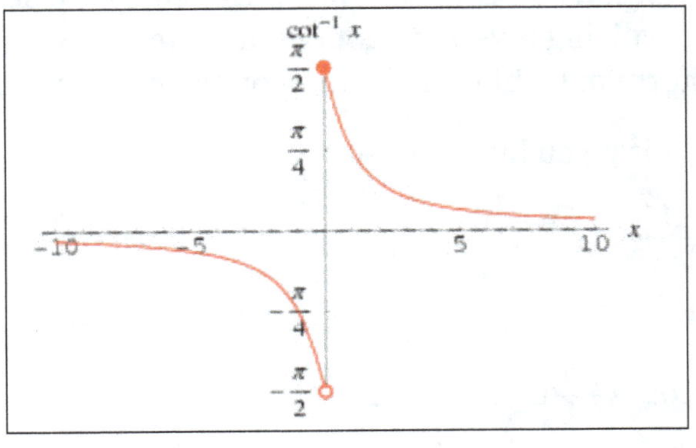

The inverse cotangent is the multivalued function $\cot^{-1} z$, also denoted arccot z or arcctg z, that is the inverse function of the cotangent. The variants Arccot z and $\mathrm{Cot}^{-1} z$ are sometimes used to refer to explicit principal values of the inverse cotangent, although this distinction is not always made. Worse yet, the notation arccot z is sometimes used for the principal value, with Arccotz being used for the multivalued function. Note that in the notation $\cot^{-1} z$ (commonly used in North America and in pocket calculators worldwide), cot z is the cotangent and the superscript -1 denotes an inverse function, not the multiplicative inverse.

The principal value of the inverse cotangent is implemented in the Wolfram Language as arccot z.

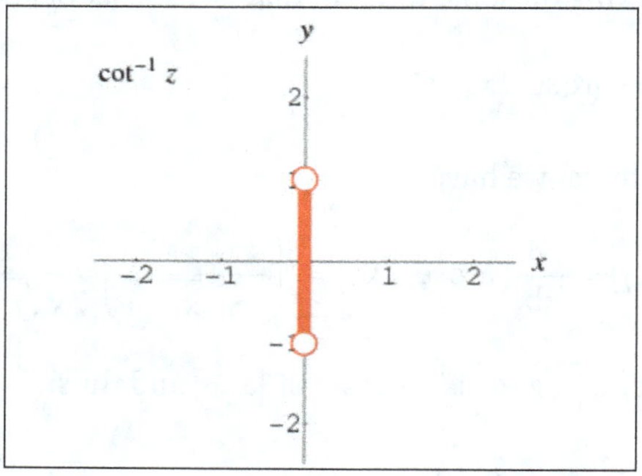

There are at least two possible conventions for defining the inverse cotangent. Taking $\cot^{-1} x$ to have range $(-\pi/2, \pi/2]$, a discontinuity at x = 0, and the branch cut placed along the line segment (-i, i). This definition can be expressed in terms of the natural logarithm by:

$$\cot^{-1} z = \frac{i}{2}\left[\mathrm{In}\left(\frac{z-i}{z}\right) - \ln\left(\frac{z+i}{z}\right)\right]$$

A different but common convention defines the range of $\cot^{-1} x$ as $(0, \pi)$, thus giving a function that is continuous on the real line R. Extreme care should be taken where examining identities involving inverse trigonometric functions, since their range of applicability or precise form may differ depending on the convention being used.

The derivative of $\cot^{-1} z$ is given by:

$$\frac{d}{dz}\cot^{-1} z = -\frac{1}{1+z^2},$$

and the integral by:

$$\int \cot^{-1} z \, dz = z\cot^{-1} z + \frac{1}{2}\ln(1+z^2) + C.$$

The Maclaurin series of the inverse cotangent for x > 0 is given by:

$$\cot^{-1}x = \frac{\pi}{2} - \sum_{k=0}^{\infty}\frac{(-1)^k x^{-(2k+1)}}{2k+1}$$

$$= \frac{\pi}{2} - x + \frac{1}{3}x^3 - \frac{1}{5}x^5 + \frac{1}{7}x^7 - \frac{1}{9}x^9 + \ldots$$

The Laurent series about z = ∞ is given by:

$$\cot^{-1}z = \sum_{k=0}^{\infty}\frac{(-1)^k z^{-(2k+1)}}{2k+1} = z^{-1} - \frac{1}{3}z^{-3} + \frac{1}{5}z^{-5} - \frac{1}{7}z^{-7} + \frac{1}{9}z^{-9} + \ldots$$

For $|z| > 1$,

Euler derived the infinite series:

$$\cot^{-1}z = z\sum_{n=1}^{\infty}\frac{(2n-2)!}{(2n-1)!!\left(z^2+1\right)^n}.$$

The inverse cotangent satisfies:

$$\cot^{-1}z = \tan^{-1}\left(\frac{1}{z}\right).$$

For $z \neq 0$,

$$\cot^{-1}z = -\cot^{-1}(-z).$$

For all $z \in \mathbb{C}^*$, and

$$\cot^{-1}x = \begin{cases} \sec^{-1}\left(\dfrac{\sqrt{x^2+1}}{x}\right) - \pi & \text{for } x<0 \\ \sec^{-1}\left(\dfrac{\sqrt{x^2+1}}{x}\right) & \text{for } x>0 \end{cases}$$

$$= \begin{cases} -\dfrac{1}{2}\pi - \tan^{-1}x & \text{for } x<0 \\ \dfrac{1}{2}\pi - \tan^{-1}x & \text{for } x\geq 0 \end{cases}$$

$$= \begin{cases} -\sin^{-1}\left(\dfrac{1}{\sqrt{x^2+1}}\right) & \text{for } x<0 \\ \sin^{-1}\left(\dfrac{1}{\sqrt{x^2+1}}\right) & \text{for } x>0 \end{cases}$$

$$= \begin{cases} -\frac{1}{2}\pi - \cot^{-1}\left(\frac{1}{x}\right) & \text{for } x<0 \\ \frac{1}{2}\pi - \cot^{-1}\left(\frac{1}{x}\right) & \text{for } x>0 \end{cases}$$

$$= \begin{cases} -\csc^{-1}\left(\sqrt{x^2+1}\right) & \text{for } x<0 \\ \csc^{-1}\left(\sqrt{x^2+1}\right) & \text{for } x>0 \end{cases}$$

$$= \begin{cases} \cos^{-1}\left(\frac{x}{\sqrt{x^2+1}}\right) - \pi & \text{for } x<0 \\ \cos^{-1}\left(\frac{x}{\sqrt{x^2+1}}\right) & \text{for } x>0 \end{cases}$$

$$= \begin{cases} -\frac{1}{2}\pi - \sin^{-1}\left(\frac{x}{\sqrt{x^2+1}}\right) & \text{for } x<0 \\ \frac{1}{2}\pi - \sin^{-1}\left(\frac{x}{\sqrt{x^2+1}}\right) & \text{for } x>0. \end{cases}$$

Analytic sums of cotangents include the beautiful result:

$$\sum_{n=1}^{i} \cot^{-1} n^2 = \cot^{-1}\left(\frac{1+t}{1-t}\right) = 1.42474...,$$

where,

$$t \equiv \cot\left(\frac{1}{2}\pi\sqrt{2}\right) \tanh\left(\frac{1}{2}\pi\sqrt{2}\right)$$

A number,

$$t_x = \cot^{(-1)} x,$$

where x is an integer or rational number, is sometimes called a Gregory number. Lehmer showed that $\cot^{-1}(a/b)$ can be expressed as a finite sum of inverse cotangents of integer arguments,

$$\cot^{-1}\left(\frac{a}{b}\right) = \sum_{i=1}^{k} (-1)^{i-1} \cot^{-1} n_i,$$

where,

$$n_i = \left|\frac{a_i}{b_i}\right|.$$

With |x| the floor function, and

$$a_{i+1} = a_i\, n + i + b_i$$
$$b_{i+1} = a_i - n_i b_i,$$

With a_o = a and b_o = b, and where the recurrence is continued until $b_{(k+1)}$ = 0. If an inverse tangent sum is written as:

$$\tan^{-1} n = \sum_{k=1} f_k \tan^{-1} n_k + f \tan^{-1} 1$$

Then equation (◇) becomes:

$$\cot^{-1} n = \sum_{k=1} f_k \cot^{-1} n_k + c \cot^{-1} 1$$

Where,

$$c = 2 - f - 2\sum_{k=1} f_k$$

Inverse cotangent sums can be used to generate Machin-like formulas.

Other inverse cotangent identities include:

$$2\cot^{-1}(2x) - \cot^{-1} x = \cot^{-1}(4x^3 + 3x)$$

$$3\cot^{-1}(3x) - \cot^{-1} x = \cot^{-1}\left(\frac{27x^4 + 18x^2 - 1}{8x}\right),$$

as well as many others. Note that for equation (previous), the choice of convention for $\cot^{-1} z$ is significant, since it holds for all complex z in the [0, π] convention, but holds only outside a lens-shaped region centered on the origin in the [-π/2, π/2] convention.

PROOFS OF DERIVATIVES OF INVERSE TRIGONOMETRIC FUNCTIONS

The following derivatives are found by setting a variable y equal to the inverse trigonometric function that we wish to take the derivative of. Using implicit differentiation and then solving for dy/dx, the derivative of the inverse function is found in terms of y. To convert dy/dx back into being in terms of x, we can draw a reference triangle on the unit circle, letting θ be y. Using the Pythagorean theorem and the definition of the regular trigonometric functions, we can finally express dy/dx in terms of x.

Differentiating the Inverse Sine Function

We let,

$$y = \arcsin x$$

Where,

$$-\frac{\pi}{2} \leq y \leq \frac{\pi}{2}$$

Then,

$$\sin y = x$$

Using implicit differentiation and solving for dy/dx:

$$\frac{d}{dx}\sin y = \frac{d}{dx}x$$

$$\frac{dy}{dx}\cos y = 1$$

Substituting $\cos y = \sqrt{1-\sin^2 y}$ in from above,

$$\frac{dy}{dx}\sqrt{1-\sin^2 y} = 1$$

Substituting $x = \sin y$ in from above,

$$\frac{dy}{dx}\sqrt{1-x^2} = 1$$

$$\frac{dy}{dx} = \frac{1}{\sqrt{1-x^2}}$$

Differentiating the Inverse Cosine Function

We let,

$$y = \arccos x$$

Where,

$$0 \leq y \leq \pi$$

Then,

$$\cos y = x$$

Using implicit differentiation and solving for dy/dx:

$$\frac{d}{dx}\cos y = \frac{d}{dx}x$$

$$-\frac{dy}{dx}\sin y = 1$$

Substituting $\sin y = \sqrt{1-\cos^2 y}$ in from above, we get:

$$-\frac{dy}{dx}\sqrt{1-\cos^2 y} = 1$$

Substituting $x = \cos y$ in from above, we get:

$$-\frac{dy}{dx}\sqrt{1-x^2} = 1$$

$$\frac{dy}{dx} = -\frac{1}{\sqrt{1-x^2}}$$

Differentiating the Inverse Tangent Function

We let:

$$y = \arctan x$$

Where,

$$-\frac{\pi}{2} < y < \frac{\pi}{2}$$

Then,

$$\tan y = x$$

Using implicit differentiation and solving for dy/dx:

$$\frac{d}{dx}\tan y = \frac{d}{dx}x$$

Left side:

$$\frac{d}{dx}\tan y = \frac{d}{dx}\frac{\sin y}{\cos y} = \frac{\frac{dy}{dx}\cos^2 y + \sin^2 y\frac{dy}{dx}}{\cos^2 y} = \frac{dy}{dx}\left(1+\tan^2 y\right)$$

Right side:

$$\frac{d}{dx}x = 1$$

Therefore,

$$\frac{dy}{dx}(1+\tan^2 y)=1$$

Substituting $x=\tan y$ in from above, we get:

$$\frac{dy}{dx}(1+x^2)=1$$

$$\frac{dy}{dx}=\frac{1}{1+x^2}$$

Differentiating the Inverse Cotangent Function

We let,

$$y=\operatorname{arccot} x$$

Where, $-\infty<x<\infty$ and $0<y<\pi$. Then,

$$\cot y=x$$

Taking a derivative with respect to x on both sides,

$$\frac{d}{dx}\cot y=\frac{d}{dx}x$$

$$-\csc^2 y\cdot\frac{dy}{dx}=1$$

$$\frac{dy}{dx}=-\frac{1}{\csc^2 y}$$

Substituting the Pythagorean identity $1+\cot^2 y=\csc^2 y$ and $x=\cot y$.

$$\frac{dy}{dx}=-\frac{1}{1+\cot^2 y}=-\frac{1}{1+x^2}$$

Differentiating the Inverse Secant Function

Let,

$$y=\operatorname{arcsec} x \quad |x|\geq 1$$

Then,

$$x=\sec y \quad y\in\left[0,\frac{\pi}{2}\right)\cup\left(\frac{\pi}{2},\pi\right]$$

$$\frac{dx}{dy} = \sec y \tan y = |x| \sqrt{x^2 - 1}$$

(It is necessary to use the absolute value of x such that the radical $x^2 - 1$ in the derivation is always its true, unnegated value.)

$$\frac{dy}{dx} = \frac{1}{|x| \sqrt{x^2 - 1}}$$

Differentiating the Inverse Cosecant Function

Let,

$$y = \operatorname{arccsc} x \quad |x| \geq 1$$

Then,

$$x = \csc y \quad y \in \left[-\frac{\pi}{2}, 0 \right) \cup \left(0, \frac{\pi}{2} \right]$$

$$\frac{dx}{dy} = -\csc y \cot y = -|x| \sqrt{x^2 - 1}$$

$$\frac{dy}{dx} = \frac{-1}{|x| \sqrt{x^2 - 1}}$$

RELATIONSHIPS AMONG THE INVERSE TRIGONOMETRIC FUNCTIONS

Basic Relations between Inverse Trigonometric Functions

$$\sin^{-1} x + \cos^{-1} x = \frac{\pi}{2} \qquad \csc^{-1} x = \sin^{-1} \frac{1}{x}$$

$$\tan^{-1} x + \cot^{-1} x = \frac{\pi}{2} \qquad \sec^{-1} x = \cos^{-1} \frac{1}{x}$$

$$\sec^{-1} x + \csc^{-1} x = \frac{\pi}{2} \qquad \cot^{-1} x = \tan^{-1} \frac{1}{x}$$

Negative Argument Formulas

$$\sin^{-1}(-x) = -\sin^{-1} x \qquad \sec^{-1}(-x) = \pi - \sec^{-1} x$$

$$\cos^{-1}(-x) = \pi - \cos^{-1} x \qquad \csc^{-1}(-x) = -\csc^{-1} x$$

$$\tan^{-1}(-x) = -\tan^{-1} x \qquad \cot^{-1}(-x) = \pi - \cot^{-1} x$$

References

- Inverse-trigonometric-formulas, maths-inverse-trigonometric-functions: toppr.com, Retrieved 14 June, 2020
- Arcsin-of-infinity, math-trigonometry-arcsin: rapidtables.com, Retrieved 25 January, 2020
- Properties-inverse-functions, maths-inverse-trigonometric-functions: toppr.com, Retrieved 18 May, 2020
- InverseSine: mathworld.wolfram.com, Retrieved 13 July, 2020
- Arc-tangent-function: calculus.subwiki.org, Retrieved 17 August, 2020
- Inverse-trig-relation, math-trig-functions: efunda.com, Retrieved 09 July, 2020

4

Trigonometric Identities and Formulas

Hermite's cotangent identity, Morrie's law, Pythagorean identities, double angle formulas, half-angle formulas, triple-angle formulas of sine, cosine and tangent, etc. are studied within trigonometric identities. This chapter closely examines these trigonometric identities and formulas to provide an extensive understanding of the subject.

All these trig identities can be derived from first principles. But there are a lot of them and some are hard to remember.

Reciprocal Identities

$$\sin A = \frac{1}{\csc A} \quad \csc A = \frac{1}{\sin A}$$

$$\cos A = \frac{1}{\sec A} \quad \sec A = \frac{1}{\cos A}$$

$$\tan A = \frac{1}{\cot A} \quad \cot A = \frac{1}{\tan A}$$

Ratio Identities

$$\tan A = \frac{\sin A}{\cos A}$$

$$\cot A = \frac{\cos A}{\sin A}$$

Opposite Angle Identities

$$\sin(-A) = -\sin A$$

$$\cos(-A) = +\cos A$$

$$\tan(-A) = -\tan A$$

Pythagorean Identities

$$\sin^2 A + \cos^2 A = 1$$
$$1 + \tan^2 A = \sec^2 A$$
$$1 + \cot^2 A = \csc^2 A$$

Complementary Angle Identities

$$\sin A = \cos\left(\frac{\pi}{2} - A\right)$$
$$\tan A = \cot\left(\frac{\pi}{2} - A\right)$$
$$\sec A = \csc\left(\frac{\pi}{2} - A\right)$$

Note: $\frac{\pi}{2}$ is 90° in radians.

If A is in degrees, use 90 instead of $\frac{\pi}{2}$.

For example: sin A = cos(90–A)

Supplementary Angle Identities

This basically says that if two angles are supplementary (add to 180°) they have the same sine.'

$$\sin A = \sin(\pi - A); \quad (0 <= A <= \pi)$$
or in degrees:
$$\sin A = \sin(180 - A); \quad (0 <= A <= 180)$$

The Sum Identities

$$\sin(A+B) = \sin A \cos B + \cos A \sin B$$
$$\cos(A+B) = \cos A \cos B - \sin A \sin B$$
$$\tan(A+B) = \frac{\tan A + \tan B}{1 - \tan A \tan B}$$

The Difference Identities

$$\sin(A-B) = \sin A \cos B - \cos A \sin B$$
$$\cos(A-B) = \cos A \cos B + \sin A \sin B$$

$$\tan(A-B)=\frac{\tan A-\tan B}{1-\tan A\tan B}$$

The Double Angle Identities

$$\sin 2A = 2\sin A\cos A$$

$$\cos 2A = \cos^2 A - \sin^2 A$$

$$\cos 2A = 2\cos^2 A - 1$$

$$\cos 2A = 1 - 2\sin^2 A$$

$$\tan 2A = \frac{2\tan A}{1-\tan^2 A}$$

The Half Angle Identities

$$\sin\frac{A}{2} = \sqrt{\frac{1-\cos A}{2}}$$

$$\cos\frac{A}{2} = \sqrt{\frac{1+\cos A}{2}}$$

$$\tan\frac{A}{2} = \frac{\sin A}{1+\cos A}$$

$$\tan\frac{A}{2} = \frac{1-\cos A}{\sin A}$$

The Sine Identities

These show how to represent the sine function in terms of the other five functions. Some of these identities may also appear under other headings.

$$\sin A = \pm\sqrt{1-\cos^2 A}$$

$$\sin A = \pm\frac{\tan A}{\sqrt{1+\tan^2 A}}$$

$$\sin A = \frac{1}{\csc A}$$

$$\sin A = \pm\frac{\sqrt{\sec^2 A - 1}}{\sec A}$$

$$\sin A = \pm\frac{1}{\sqrt{1+\cot^2 A}}$$

The Cosine Identities

These show how to represent the cosine function in terms of the other five functions. Some of these identities may also appear under other headings.

$$\cos A = \pm\sqrt{1-\sin^2 A}$$

$$\cos A = \pm\frac{1}{\sqrt{1+\tan^2 A}}$$

$$\cos A = \frac{1}{\sec A}$$

$$\cos A = \pm\frac{\sqrt{\csc^2 A-1}}{\csc A}$$

$$\cos A = \pm\frac{\cot A}{\sqrt{1+\cot^2 A}}$$

The Tangent Identities

These show how to represent the tangent function in terms of the other five functions. Some of these trig identities may also appear under other headings.

$$\tan A = \pm\sqrt{\sec^2 - 1}$$

$$\tan A = \pm\frac{1}{\sqrt{\csc^2 A-1}}$$

$$\tan A = \frac{1}{\cot A}$$

$$\tan A = \pm\frac{\sqrt{1-\cos^2 A}}{\cos A}$$

$$\tan A = \pm\frac{\sin A}{\sqrt{1-\sin^2 A}}$$

The Cosecant Identities

These show how to represent the cosecant function in terms of the other five functions. Some of these trig identities may also appear under other headings.

$$\csc A = \pm\sqrt{1-\cot^2 A}$$

$$\csc A = \pm\frac{1}{\sqrt{1-\cos^2 A}}$$

$$\csc A = \frac{1}{\sin A}$$

$$\csc A = \pm\frac{\sqrt{1+\tan^2 A}}{\tan A}$$

$$\csc A = \pm\frac{\sin A}{\sqrt{\sec^2 A - 1}}$$

The Secant Identities

These show how to represent the secant function in terms of the other five functions. Some of these trig identities may also appear under other headings.

$$\sec A = \pm\sqrt{1+\tan^2 A}$$

$$\sec A = \pm\frac{1}{\sqrt{1-\sin^2 A}}$$

$$\sec A = \frac{1}{\cos A}$$

$$\sec A = \pm\frac{\sqrt{1+\cot^2 A}}{\cot A}$$

$$\sec A = \pm\frac{\csc A}{\sqrt{\csc^2 A - 1}}$$

The Cotangent Identities

These show how to represent the cotangent function in terms of the other five functions. Some of these identities may also appear under other headings.

$$\cot A = \pm\sqrt{\csc^2 - 1}$$

$$\cot A = \pm\frac{1}{\sqrt{\sec^2 A - 1}}$$

$$\cot A = \frac{1}{\tan A}$$

$$\cot A = \pm\frac{\sqrt{1-\sin^2 A}}{\sin A}$$

$$\cot A = \pm\frac{\cos A}{\sqrt{1-\cos^2 A}}$$

Trigonometric Constants Expressed in Real Radicals

Exact algebraic expressions for trigonometric values are sometimes useful, mainly for simplifying solutions into radical forms which allow further simplification.

All trigonometric numbers – sines or cosines of rational multiples of 360° – are algebraic numbers (solutions of polynomial equations with integer coefficients); moreover they may be expressed in terms of radicals of complex numbers; but not all of these are expressible in terms of *real* radicals. When they are, they are expressible more specifically in terms of square roots.

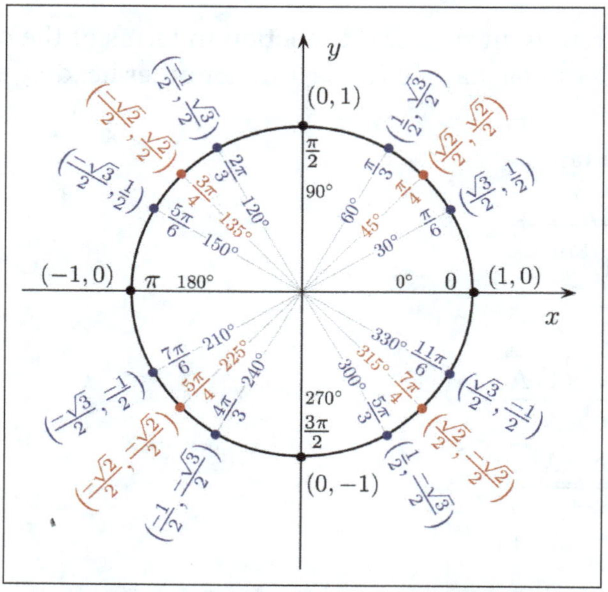

The primary solution angles in the form (cos,sin) on the unit circle are at multiples of 30 and 45 degrees.

All values of the sines, cosines, and tangents of angles at 3° increments are expressible in terms of square roots, using identities – the half-angle identity, the double-angle identity, and the angle addition/subtraction identity – and using values for 0°, 30°, 36°, and 45°. For an angle of an integer number of degrees that is not a multiple 3° ($\frac{\pi}{60}$ radians), the values of sine, cosine, and tangent cannot be expressed in terms of real radicals.

According to Niven's theorem, the only rational values of the sine function for which the argument is a rational number of degrees are 0, $\frac{1}{2}$, 1, $-\frac{1}{2}$, and −1.

According to Baker's theorem, if the value of a sine, a cosine or a tangent is algebraic, then the angle is either a rational number of degrees or a transcendental number of degrees. That is, if the angle is an algebraic, but non-rational, number of degrees, the trigonometric functions all have transcendental values.

Table of some Common Angles

Several different units of angle measure are widely used, including degrees, radians, and gradians (gons):

1 full circle (turn) = 360 degrees = 2π radians = 400 gons.

The following table shows the conversions and values for some common angles:

Turns	Degrees	Radians	Gradians	sine	cosine	tangent
0	$0°$	0	0^g	0	1	0
$\frac{1}{12}$	$30°$	$\frac{\pi}{6}$	$33\frac{1}{3}^g$	$\frac{1}{2}$	$\frac{\sqrt{3}}{2}$	$\frac{\sqrt{3}}{3}$
$\frac{1}{8}$	$45°$	$\frac{\pi}{4}$	50^g	$\frac{\sqrt{2}}{2}$	$\frac{\sqrt{2}}{2}$	1
$\frac{1}{6}$	$60°$	$\frac{\pi}{3}$	$66\frac{2}{3}^g$	$\frac{\sqrt{3}}{2}$	$\frac{1}{2}$	$\sqrt{3}$
$\frac{1}{4}$	$90°$	$\frac{\pi}{2}$	100^g	1	0	
$\frac{1}{3}$	$120°$	$\frac{2\pi}{3}$	$133\frac{1}{3}^g$	$\frac{\sqrt{3}}{2}$	$-\frac{1}{2}$	$-\sqrt{3}$
$\frac{3}{8}$	$135°$	$\frac{3\pi}{4}$	150^g	$\frac{\sqrt{2}}{2}$	$-\frac{\sqrt{2}}{2}$	-1
$\frac{5}{12}$	$150°$	$\frac{5\pi}{6}$	$166\frac{2}{3}^g$	$\frac{1}{2}$	$-\frac{\sqrt{3}}{2}$	$-\frac{\sqrt{3}}{3}$
$\frac{1}{2}$	$180°$	π	200^g	0	-1	0
$\frac{7}{12}$	$210°$	$\frac{7\pi}{6}$	$233\frac{1}{3}^g$	$-\frac{1}{2}$	$-\frac{\sqrt{3}}{2}$	$\frac{\sqrt{3}}{3}$
$\frac{5}{8}$	$225°$	$\frac{5\pi}{4}$	250^g	$-\frac{\sqrt{2}}{2}$	$-\frac{\sqrt{2}}{2}$	1
$\frac{2}{3}$	$240°$	$\frac{4\pi}{3}$	$266\frac{2}{3}^g$	$-\frac{\sqrt{3}}{2}$	$-\frac{1}{2}$	$\sqrt{3}$
$\frac{3}{4}$	$270°$	$\frac{3\pi}{2}$	300^g	-1	0	
$\frac{5}{8}$	$300°$	$\frac{5\pi}{3}$	$333\frac{1}{3}^g$	$-\frac{\sqrt{3}}{2}$	$\frac{1}{2}$	$-\sqrt{3}$
$\frac{7}{8}$	$315°$	$\frac{7\pi}{4}$	350^g	$-\frac{\sqrt{2}}{2}$	$\frac{\sqrt{2}}{2}$	-1

$\frac{11}{2}$	330°	$\frac{11\pi}{6}$	$366\frac{2}{3}^{g}$	$-\frac{1}{2}$	$\frac{\sqrt{3}}{2}$	$-\frac{\sqrt{3}}{3}$
1	360°	2π	400^{g}	0	1	0

Further Angles

Table: Exact trigonometric table for multiples of 3 degrees.

θ	cos θ	sin θ
0°	1	0
3°	$\frac{2\left(\sqrt{3}+1\right)\sqrt{5+\sqrt{5}}+\sqrt{2}\left(\sqrt{3}-1\right)\left(\sqrt{5}-1\right)}{16}$	$\frac{2\left(\sqrt{3}+1\right)\left(\sqrt{5}-1\right)-2\left(\sqrt{3}-1\right)\left(\sqrt{5+\sqrt{5}}\right)}{16}$
6°	$\frac{\sqrt{2}\sqrt{5-\sqrt{5}}+\sqrt{3}\left(\sqrt{5}+1\right)}{8}$	$\frac{\sqrt{6}\sqrt{5-\sqrt{5}}-\left(\sqrt{5}+1\right)}{8}$
9°	$\frac{\sqrt{2}\left(\sqrt{5}+1\right)+\left(\sqrt{5}-1\right)\sqrt{5+\sqrt{5}}}{8}$	$\frac{\sqrt{2}\left(\sqrt{5}+1\right)-\left(\sqrt{5}-1\right)\sqrt{5+\sqrt{5}}}{8}$
12°	$\frac{\sqrt{6}\sqrt{5+\sqrt{5}}+\sqrt{5}-1}{8}$	$\frac{\sqrt{2}\sqrt{5+\sqrt{5}}-\sqrt{3}\left(\sqrt{5}-1\right)}{8}$
15°	$\frac{\sqrt{2+\sqrt{3}}}{2}$	$\frac{\sqrt{2-\sqrt{3}}}{2}$
18°	$\frac{\sqrt{2}\sqrt{5+\sqrt{5}}}{4}$	$\frac{\sqrt{5}-1}{4}$
21°	$\frac{2\left(\sqrt{3}-1\right)\sqrt{5-\sqrt{5}}+\sqrt{2}\left(\sqrt{3}+1\right)\left(\sqrt{5}+1\right)}{16}$	$\frac{2\left(\sqrt{3}+1\right)\sqrt{5-\sqrt{5}}-\sqrt{2}\left(\sqrt{3}-1\right)\left(\sqrt{5}+1\right)}{16}$
24°	$\frac{\sqrt{5}+1+\sqrt{6}\sqrt{5-\sqrt{5}}}{8}$	$\frac{\sqrt{3}\left(\sqrt{5}+1\right)-\sqrt{2}\sqrt{5-\sqrt{5}}}{8}$
27°	$\frac{2\sqrt{5+\sqrt{5}}+\sqrt{2}\left(\sqrt{5}-1\right)}{8}$	$\frac{2\sqrt{5+\sqrt{5}}-\sqrt{2}\left(\sqrt{5}-1\right)}{8}$
30°	$\frac{\sqrt{3}}{2}$	$\frac{1}{2}$
33°	$\frac{2\left(\sqrt{3}+1\right)\sqrt{5+\sqrt{5}}-\sqrt{2}\left(\sqrt{3}-1\right)\left(\sqrt{5}-1\right)}{16}$	$\frac{2\left(\sqrt{3}-1\right)\sqrt{5+\sqrt{5}}+\sqrt{2}\left(\sqrt{3}+1\right)\left(\sqrt{5}-1\right)}{16}$
36°	$\frac{\sqrt{5}+1}{4}$	$\frac{\sqrt{2}\sqrt{5-\sqrt{5}}}{4}$

39°	$\frac{\sqrt{2}\left(\sqrt{3}-1\right)\left(\sqrt{5}+1\right)+2\left(\sqrt{3}+1\right)\sqrt{5-\sqrt{5}}}{16}$	$\frac{\sqrt{2}\left(\sqrt{3}+1\right)\left(\sqrt{5}+1\right)-2\left(\sqrt{3}-1\right)\sqrt{5-\sqrt{5}}}{16}$
42°	$\frac{\sqrt{2}\sqrt{5+\sqrt{5}}+\sqrt{3}\left(\sqrt{5}-1\right)}{8}$	$\frac{\sqrt{6}\sqrt{5+\sqrt{5}}-\left(\sqrt{5}-1\right)}{8}$
45°	$\frac{\sqrt{2}}{2}$	$\frac{\sqrt{2}}{2}$

Values outside the [0°, 45°] angle range are trivially derived from these values, using circle axis reflection symmetry.

In the entries below, when a certain number of degrees is related to a regular polygon, the relation is that the number of degrees in each angle of the polygon is $(n - 2)$ times the indicated number of degrees (where n is the number of sides). This is because the sum of the angles of any n-gon is $180° \times (n - 2)$ and so the measure of each angle of any regular n-gon is $180° \times (n - 2) \div n$. Thus for example the entry "45°: square" means that, with $n = 4$, $180° \div n = 45°$, and the number of degrees in each angle of a square is $(n - 2) \times 45° = 90°$.

0°: Fundamental

$$\sin 0 = 0$$

$$\cos 0 = 1$$

$$\tan 0 = 0$$

$\cot 0$ is undefined

1.5°: Regular Hecatonicosagon (120-sided Polygon)

$$\sin\left(\frac{\pi}{120}\right)=\sin\left(1.5°\right)=\frac{\left(\sqrt{2+\sqrt{2}}\right)\left(\sqrt{15}+\sqrt{3}-\sqrt{10-2\sqrt{5}}\right)-\left(\sqrt{2-\sqrt{2}}\right)\left(\sqrt{30-6\sqrt{5}}+\sqrt{5}+1\right)}{16}$$

$$\cos\left(\frac{\pi}{120}\right)=\cos\left(1.5°\right)=\frac{\left(\sqrt{2+\sqrt{2}}\right)\left(\sqrt{30-6\sqrt{5}}+\sqrt{5}+1\right)+\left(\sqrt{2-\sqrt{2}}\right)\left(\sqrt{15}+\sqrt{3}-\sqrt{10-2\sqrt{5}}\right)}{16}$$

1.875°: Regular Enneacontahexagon (96-sided Polygon)

$$\sin\left(\frac{\pi}{96}\right)=\sin\left(1.875°\right)=\frac{1}{2}\sqrt{2-\sqrt{2+\sqrt{2+\sqrt{2+\sqrt{3}}}}}$$

$$\cos\left(\frac{}{96}\right)=\cos\left(1.875°\right)=\frac{}{2}\sqrt{2+\sqrt{2+\sqrt{2+\sqrt{2+\sqrt{3}}}}}$$

2.25°: Regular Octacontagon (80-sided Polygon)

$$\sin\left(\frac{\pi}{80}\right)=\sin(2.25^\circ)=\frac{1}{2}\sqrt{2-\sqrt{2+\sqrt{2+\sqrt{\frac{5+\sqrt{5}}{2}}}}}$$

$$\cos\left(\frac{\pi}{80}\right)=\cos(2.25^\circ)=\frac{1}{2}\sqrt{2+\sqrt{2+\sqrt{2+\sqrt{\frac{5+\sqrt{5}}{2}}}}}$$

2.8125°: Regular Hexacontatetragon (64-sided Polygon)

$$\sin\left(\frac{\pi}{64}\right)=\sin(2.8125^\circ)=\frac{1}{2}\sqrt{2-\sqrt{2+\sqrt{2+\sqrt{2+\sqrt{2}}}}}$$

$$\cos\left(\frac{\pi}{64}\right)=\cos(2.8125^\circ)=\frac{1}{2}\sqrt{2+\sqrt{2+\sqrt{2+\sqrt{2+\sqrt{2}}}}}$$

3°: Regular Hexacontagon (60-sided Polygon)

$$\sin\left(\frac{\pi}{60}\right)=\sin(3^\circ)=\frac{2\left(1-\sqrt{3}\right)\sqrt{5+\sqrt{5}}+\left(\sqrt{10}-\sqrt{2}\right)\left(\sqrt{3}+1\right)}{16}$$

$$\cos\left(\frac{\pi}{60}\right)=\cos(3^\circ)=\frac{2\left(1+\sqrt{3}\right)\sqrt{5+\sqrt{5}}+\left(\sqrt{10}-\sqrt{2}\right)\left(\sqrt{3}-1\right)}{16}$$

$$\tan\left(\frac{\pi}{60}\right)=\tan(3^\circ)=\frac{\left[\left(2-\sqrt{3}\right)\left(3+\sqrt{5}\right)-2\right]\left[2-\sqrt{10-2\sqrt{5}}\right]}{4}$$

$$\cot\left(\frac{\pi}{60}\right)=\cot(3^\circ)=\frac{\left[\left(2+\sqrt{3}\right)\left(3+\sqrt{5}\right)-2\right]\left[2+\sqrt{10-2\sqrt{5}}\right]}{4}$$

3.75°: Regular Tetracontaoctagon (48-sided Polygon)

$$\sin\left(\frac{\pi}{48}\right)=\sin(3.75^\circ)=\frac{1}{2}\sqrt{2-\sqrt{2+\sqrt{2+\sqrt{3}}}}$$

$$\cos\left(\frac{\pi}{48}\right)=\cos(3.75^\circ)=\frac{1}{2}\sqrt{2+\sqrt{2+\sqrt{2+\sqrt{3}}}}$$

4.5°: Regular Tetracontagon (40-sided Polygon)

$$\sin\left(\frac{\pi}{40}\right)=\sin(4.5°)=\frac{1}{2}\sqrt{2-\sqrt{2+\sqrt{\frac{5+\sqrt{5}}{2}}}}$$

$$\cos\left(\frac{\pi}{40}\right)=\cos(4.5°)=\frac{1}{2}\sqrt{2+\sqrt{2+\sqrt{\frac{5+\sqrt{5}}{2}}}}$$

5.625°: Regular Triacontadigon (32-sided Polygon)

$$\sin\left(\frac{\pi}{32}\right)=\sin(5.625°)=\frac{1}{2}\sqrt{2-\sqrt{2+\sqrt{2+\sqrt{2}}}}$$

$$\cos\left(\frac{\pi}{32}\right)=\cos(5.625°)=\frac{1}{2}\sqrt{2+\sqrt{2+\sqrt{2+\sqrt{2}}}}$$

6°: Regular Triacontagon (30-sided Polygon)

$$\sin\frac{\pi}{30}=\sin 6°=\frac{\sqrt{30-\sqrt{180}}-\sqrt{5}-1}{8}$$

$$\cos\frac{\pi}{30}=\cos 6°=\frac{\sqrt{10-\sqrt{20}}+\sqrt{3}+\sqrt{15}}{8}$$

$$\tan\frac{\pi}{30}=\tan 6°=\frac{\sqrt{10-\sqrt{20}}+\sqrt{3}-\sqrt{15}}{2}$$

$$\cot\frac{\pi}{30}=\cot 6°=\frac{\sqrt{27}+\sqrt{15}+\sqrt{50+\sqrt{2420}}}{2}$$

7.5°: Regular Icositetragon (24-sided Polygon)

$$\sin\left(\frac{\pi}{24}\right)=\sin(7.5°)=\frac{1}{2}\sqrt{2-\sqrt{2+\sqrt{3}}}=\frac{1}{4}\sqrt{8-2\sqrt{6}-2\sqrt{2}}$$

$$\cos\left(\frac{\pi}{24}\right)=\cos(7.5°)=\frac{1}{2}\sqrt{2+\sqrt{2+\sqrt{3}}}=\frac{1}{4}\sqrt{8+2\sqrt{6}+2\sqrt{2}}$$

$$\tan\left(\frac{\pi}{24}\right)=\tan(7.5°)=\sqrt{6}-\sqrt{3}+\sqrt{2}-2=\left(\sqrt{2}-1\right)\left(\sqrt{3}-\sqrt{2}\right)$$

$$\cot\left(\frac{\pi}{24}\right) = \cot(7.5°) = \sqrt{6} + \sqrt{3} + \sqrt{2} + 2 = \left(\sqrt{2} + 1\right)\left(\sqrt{3} + \sqrt{2}\right)$$

9°: Regular Icosagon (20-sided Polygon)

$$\sin\frac{\pi}{20} = \sin 9° = \frac{1}{2}\sqrt{2 - \sqrt{\frac{5+\sqrt{5}}{2}}}$$

$$\cos\frac{\pi}{20} = \cos 9° = \frac{1}{2}\sqrt{2 + \sqrt{\frac{5+\sqrt{5}}{2}}}$$

$$\tan\frac{\pi}{20} = \tan 9° = \sqrt{5} + 1 - \sqrt{5 + 2\sqrt{5}}$$

$$\cot\frac{\pi}{20} = \cot 9° = \sqrt{5} + 1 + \sqrt{5 + 2\sqrt{5}}$$

11.25°: Regular Hexadecagon (16-sided Polygon)

$$\sin\frac{\pi}{16} = \sin 11.25° = \frac{1}{2}\sqrt{2 - \sqrt{2 + \sqrt{2}}}$$

$$\cos\frac{\pi}{16} = \cos 11.25° = \frac{1}{2}\sqrt{2 + \sqrt{2 + \sqrt{2}}}$$

$$\tan\frac{\pi}{16} = \tan 11.25° = \sqrt{4 + 2\sqrt{2}} - \sqrt{2} - 1$$

$$\cot\frac{\pi}{16} = \cot 11.25° = \sqrt{4 + 2\sqrt{2}} + \sqrt{2} + 1$$

12°: Regular Pentadecagon (15-sided Polygon)

$$\sin\frac{\pi}{15} = \sin 12° = \tfrac{1}{8}\left[\sqrt{2\left(5 + \sqrt{5}\right)} + \sqrt{3} - \sqrt{15}\right]$$

$$\cos\frac{\pi}{15} = \cos 12° = \tfrac{1}{8}\left[\sqrt{6\left(5 + \sqrt{5}\right)} + \sqrt{5} - 1\right]$$

$$\tan\frac{\pi}{15} = \tan 12° = \tfrac{1}{2}\left[3\sqrt{3} - \sqrt{15} - \sqrt{2\left(25 - 11\sqrt{5}\right)}\right]$$

$$\cot\frac{\pi}{15} = \cot 12° = \tfrac{1}{2}\left[\sqrt{15} + \sqrt{3} + \sqrt{2\left(5 + \sqrt{5}\right)}\right]$$

15°: Regular Dodecagon (12-sided Polygon)

$$\sin\frac{\pi}{12} = \sin 15^\circ = \frac{1}{4}\left(\sqrt{6} - \sqrt{2}\right) = \frac{1}{2}\sqrt{2-\sqrt{3}}$$

$$\cos\frac{\pi}{12} = \cos 15^\circ = \frac{1}{4}\left(\sqrt{6} + \sqrt{2}\right) = \frac{1}{2}\sqrt{2+\sqrt{3}}$$

$$\tan\frac{\pi}{12} = \tan 15^\circ = 2 - \sqrt{3}$$

$$\cot\frac{\pi}{12} = \cot 15^\circ = 2 + \sqrt{3}$$

18°: Regular Decagon (10-sided Polygon)

$$\sin\frac{\pi}{10} = \sin 18^\circ = \tfrac{1}{4}\left(\sqrt{5} - 1\right)$$

$$\cos\frac{\pi}{10} = \cos 18^\circ = \tfrac{1}{4}\sqrt{2\left(5+\sqrt{5}\right)}$$

$$\tan\frac{}{10} = \tan 18^\circ = -\sqrt{5\left(5-2\sqrt{5}\right)}$$

$$\cot\frac{\pi}{10} = \cot 18^\circ = \sqrt{5+2\sqrt{5}}$$

21°: Sum 9° + 12°

$$\sin\frac{7\pi}{60} = \sin 21^\circ = \frac{1}{16}\left(2\left(\sqrt{3}+1\right)\sqrt{5-\sqrt{5}} - \left(\sqrt{6}-\sqrt{2}\right)\left(1+\sqrt{5}\right)\right)$$

$$\cos\frac{7\pi}{60} = \cos 21^\circ = \frac{1}{16}\left(2\left(\sqrt{3}-1\right)\sqrt{5-\sqrt{5}} + \left(\sqrt{6}+\sqrt{2}\right)\left(1+\sqrt{5}\right)\right)$$

$$\tan\frac{7\pi}{60} = \tan 21^\circ = \frac{1}{4}\left(2-\left(2+\sqrt{3}\right)\left(3-\sqrt{5}\right)\right)\left(2-\sqrt{2\left(5+\sqrt{5}\right)}\right)$$

$$\cot\frac{7\pi}{60} = \cot 21^\circ = \frac{1}{4}\left(2-\left(2-\sqrt{3}\right)\left(3-\sqrt{5}\right)\right)\left(2+\sqrt{2\left(5+\sqrt{5}\right)}\right)$$

22.5°: Regular Octagon

$$\sin\frac{\pi}{8} = \sin 22.5^\circ = \frac{1}{2}\sqrt{2-\sqrt{2}},$$

$$\cos\frac{\pi}{8} = \cos 22.5° = \frac{1}{2}\sqrt{2+\sqrt{2}}$$

$$\tan\frac{\pi}{8} = \tan 22.5° = \sqrt{2}-1$$

$$\cot\frac{\pi}{8} = \cot 22.5° = \sqrt{2}+1 = \delta_S, \text{ the silver ratio.}$$

24°: Sum 12° + 12°

$$\sin\frac{2\pi}{15} = \sin 24° = \tfrac{1}{8}\left[\sqrt{15}+\sqrt{3}-\sqrt{2\left(5-\sqrt{5}\right)}\right]$$

$$\cos\frac{2\pi}{15} = \cos 24° = \tfrac{1}{8}\left(\sqrt{6\left(5-\sqrt{5}\right)}+\sqrt{5}+1\right)$$

$$\tan\frac{2\pi}{15} = \tan 24° = \tfrac{1}{2}\left[\sqrt{50+22\sqrt{5}}-3\sqrt{3}-\sqrt{15}\right]$$

$$\cot\frac{2\pi}{15} = \cot 24° = \tfrac{1}{2}\left[\sqrt{15}-\sqrt{3}+\sqrt{2\left(5-\sqrt{5}\right)}\right]$$

27°: Sum 12° + 15°

$$\sin\frac{3\pi}{20} = \sin 27° = \tfrac{1}{8}\left[2\sqrt{5+\sqrt{5}}-\sqrt{2}\left(\sqrt{5}-1\right)\right]$$

$$\cos\frac{3\pi}{20} = \cos 27° = \tfrac{1}{8}\left[2\sqrt{5+\sqrt{5}}+\sqrt{2}\left(\sqrt{5}-1\right)\right]$$

$$\tan\frac{3\pi}{20} = \tan 27° = \sqrt{5}-1-\sqrt{5-2\sqrt{5}}$$

$$\cot\frac{3\pi}{20} = \cot 27° = \sqrt{5}-1+\sqrt{5-2\sqrt{5}}$$

30°: Regular Hexagon

$$\sin\frac{\pi}{6} = \sin 30° = \frac{1}{2}$$

$$\cos\frac{\pi}{6} = \cos 30° = \frac{\sqrt{3}}{2}$$

$$\tan\frac{\pi}{6} = \tan 30^\circ = \frac{\sqrt{3}}{3} = \frac{1}{\sqrt{3}}$$

$$\cot\frac{\pi}{6} = \cot 30^\circ = \sqrt{3}$$

33°: Sum 15° + 18°

$$\sin\frac{11\pi}{60} = \sin 33^\circ = \tfrac{1}{16}\left[2\left(\sqrt{3}-1\right)\sqrt{5+\sqrt{5}} + \sqrt{2}\left(1+\sqrt{3}\right)\left(\sqrt{5}-1\right)\right]$$

$$\cos\frac{11\pi}{60} = \cos 33^\circ = \tfrac{1}{16}\left[2\left(\sqrt{3}+1\right)\sqrt{5+\sqrt{5}} + \sqrt{2}\left(1-\sqrt{3}\right)\left(\sqrt{5}-1\right)\right]$$

$$\tan\frac{11\pi}{60} = \tan 33^\circ = \tfrac{1}{4}\left[2-\left(2-\sqrt{3}\right)\left(3+\sqrt{5}\right)\right]\left[2+\sqrt{2\left(5-\sqrt{5}\right)}\right]$$

$$\cot\frac{11\pi}{60} = \cot 33^\circ = \tfrac{1}{4}\left[2-\left(2+\sqrt{3}\right)\left(3+\sqrt{5}\right)\right]\left[2-\sqrt{2\left(5-\sqrt{5}\right)}\right]$$

36°: Regular Pentagon

$$\sin\frac{\pi}{5} = \sin 36^\circ = \frac{1}{4}\sqrt{10-2\sqrt{5}}$$

$$\cos\frac{\pi}{5} = \cos 36^\circ = \frac{\sqrt{5}+1}{4} = \frac{\varphi}{2},$$ where φ is the golden ratio;

$$\tan\frac{\pi}{5} = \tan 36^\circ = \sqrt{5-2\sqrt{5}}$$

$$\cot\frac{\pi}{5} = \cot 36^\circ = \frac{1}{5}\sqrt{25+10\sqrt{5}}$$

39°: Sum 18° + 21°

$$\sin\frac{13\pi}{60} = \sin 39^\circ = \tfrac{1}{16}\left[2\left(1-\sqrt{3}\right)\sqrt{5-\sqrt{5}} + \sqrt{2}\left(\sqrt{3}+1\right)\left(\sqrt{5}+1\right)\right]$$

$$\cos\frac{13\pi}{60} = \cos 39^\circ = \tfrac{1}{16}\left[2\left(1+\sqrt{3}\right)\sqrt{5-\sqrt{5}} + \sqrt{2}\left(\sqrt{3}-1\right)\left(\sqrt{5}+1\right)\right]$$

$$\tan\frac{13\pi}{60} = \tan 39^\circ = \tfrac{1}{4}\left[\left(2-\sqrt{3}\right)\left(3-\sqrt{5}\right)-2\right]\left[2-\sqrt{2\left(5+\sqrt{5}\right)}\right]$$

$$\cot\frac{13\pi}{60}=\cot 39^\circ=\tfrac{1}{4}\left[\left(2+\sqrt{3}\right)\left(3-\sqrt{5}\right)-2\right]\left[2+\sqrt{2\left(5+\sqrt{5}\right)}\right]$$

42°: Sum 21° + 21°

$$\sin\frac{7\pi}{30}=\sin 42^\circ=\frac{\sqrt{30+\sqrt{180}}-\sqrt{5}+1}{8}$$

$$\cos\frac{7\pi}{30}=\cos 42^\circ=\frac{\sqrt{15}-\sqrt{3}+\sqrt{10+\sqrt{20}}}{8}$$

$$\tan\frac{7\pi}{30}=\tan 42^\circ=\frac{\sqrt{15}+\sqrt{3}-\sqrt{10+\sqrt{20}}}{2}$$

$$\cot\frac{7\pi}{30}=\cot 42^\circ=\frac{\sqrt{50-\sqrt{2420}}+\sqrt{27}-\sqrt{15}}{2}$$

45°: Square

$$\sin\frac{\pi}{4}=\sin 45^\circ=\frac{\sqrt{2}}{2}=\frac{1}{\sqrt{2}}$$

$$\cos\frac{\pi}{4}=\cos 45^\circ=\frac{\sqrt{2}}{2}=\frac{1}{\sqrt{2}}$$

$$\tan\frac{\pi}{4}=\tan 45^\circ=1$$

$$\cot\frac{\pi}{4}=\cot 45^\circ=1$$

54°: Sum 27° + 27°

$$\sin\frac{3\pi}{10}=\sin 54^\circ=\frac{\sqrt{5}+1}{4}$$

$$\cos\frac{3\pi}{10}=\cos 54^\circ=\frac{\sqrt{10-2\sqrt{5}}}{4}$$

$$\tan\frac{3\pi}{10}=\tan 54^\circ=\frac{\sqrt{25+10\sqrt{5}}}{5}$$

$$\cot\frac{3\pi}{10} = \cot 54° = \sqrt{5-\sqrt{20}}$$

60°: Equilateral Triangle

$$\sin\frac{\pi}{3} = \sin 60° = \frac{\sqrt{3}}{2}$$

$$\cos\frac{\pi}{3} = \cos 60° = \frac{1}{2}$$

$$\tan\frac{\pi}{3} = \tan 60° = \sqrt{3}$$

$$\cot\frac{\pi}{3} = \cot 60° = \frac{\sqrt{3}}{3} = \frac{1}{\sqrt{3}}$$

67.5°: Sum 7.5° + 60°

$$\sin\frac{3\pi}{8} = \sin 67.5° = \tfrac{1}{2}\sqrt{2+\sqrt{2}}$$

$$\cos\frac{3\pi}{8} = \cos 67.5° = \tfrac{1}{2}\sqrt{2-\sqrt{2}}$$

$$\tan\frac{3\pi}{8} = \tan 67.5° = \sqrt{2}+1$$

$$\cot\frac{3\pi}{8} = \cot 67.5° = \sqrt{2}-1$$

72°: Sum 36° + 36°

$$\sin\frac{2\pi}{5} = \sin 72° = \tfrac{1}{4}\sqrt{2\left(5+\sqrt{5}\right)}$$

$$\cos\frac{2\pi}{5} = \cos 72° = \tfrac{1}{4}\left(\sqrt{5}-1\right)$$

$$\tan\frac{2\pi}{5} = \tan 72° = \sqrt{5+2\sqrt{5}}$$

$$\cot\frac{2\pi}{5} = \cot 72° = \tfrac{1}{5}\sqrt{5\left(5-2\sqrt{5}\right)}$$

75°: Sum 30° + 45°

$$\sin\frac{5\pi}{12} = \sin 75° = \tfrac{1}{4}\left(\sqrt{6}+\sqrt{2}\right)$$

$$\cos\frac{5\pi}{12} = \cos 75° = \tfrac{1}{4}\left(\sqrt{6}-\sqrt{2}\right)$$

$$\tan\frac{5\pi}{12} = \tan 75° = 2+\sqrt{3}$$

$$\cot\frac{5\pi}{12} = \cot 75° = 2-\sqrt{3}$$

90°: Fundamental

$$\sin\frac{\pi}{2} = \sin 90° = 1$$

$$\cos\frac{\pi}{2} = \cos 90° = 0$$

$$\tan\frac{\pi}{2} = \tan 90° \text{ is undefined}$$

$$\cot\frac{\pi}{2} = \cot 90° = 0$$

List of Trigonometric Constants of $\frac{2\pi}{n}$

For cube roots of non-real numbers that appear in this table, one has to take the principal value, that is the cube root with the largest real part; this largest real part is always positive. Therefore, the sums of cube roots that appear in the table are all positive real numbers.

n	$\sin\left(\frac{2\pi}{n}\right)$
1	0
2	0
3	$\frac{1}{2}\sqrt{3}$
4	1
5	$\frac{1}{4}\left(\sqrt{10+2\sqrt{5}}\right)$

6	$\frac{1}{2}\sqrt{3}$
7	$\frac{1}{2}\sqrt{\frac{1}{3}\left(7-\sqrt[3]{\frac{7+21\sqrt{-3}}{2}}-\sqrt[3]{\frac{7-21\sqrt{-3}}{2}}\right)}$
8	$\frac{1}{2}\sqrt{2}$
9	$\frac{i}{2}\left(\sqrt[3]{\frac{-1-\sqrt{-3}}{2}}-\sqrt[3]{\frac{-1+\sqrt{-3}}{2}}\right)$
10	$\frac{1}{4}\left(\sqrt{10-2\sqrt{5}}\right)$
11	
12	$\frac{1}{2}$
13	
14	$\frac{1}{24}\sqrt{3\left(112-\sqrt[3]{14336+\sqrt{-5549064193}}-\sqrt[3]{14336-\sqrt{-5549064193}}\right)}$
15	$\frac{1}{8}\left(\sqrt{15}+\sqrt{3}-\sqrt{10-2\sqrt{5}}\right)$
16	$\frac{1}{2}\left(\sqrt{2-\sqrt{2}}\right)$
17	
18	$\frac{i}{4}\left(\sqrt[3]{4-4\sqrt{-3}}-\sqrt[3]{4+4\sqrt{-3}}\right)$
19	
20	$\frac{1}{4}\left(\sqrt{5}-1\right)$
21	
22	
23	
24	$\frac{1}{4}\left(\sqrt{6}-\sqrt{2}\right)$

n	$\cos\left(\frac{2\pi}{n}\right)$
1	1
2	-1
3	$-\frac{1}{2}$
4	0

5	$\frac{1}{4}\left(\sqrt{5}-1\right)$
6	$\frac{1}{2}$
7	$\frac{1}{6}\left(-1+\sqrt[3]{\frac{7+21\sqrt{-3}}{2}}+\sqrt[3]{\frac{7-21\sqrt{-3}}{2}}\right)$
8	$\frac{1}{2}\sqrt{2}$
9	$\frac{i}{2}\left(\sqrt[3]{\frac{-1+\sqrt{-3}}{2}}-\sqrt[3]{\frac{-1-\sqrt{-3}}{2}}\right)$
10	$\frac{1}{4}\left(\sqrt{5}+1\right)$
11	
12	$\frac{1}{2}\sqrt{3}$
13	
14	$\frac{1}{24}\sqrt{3\left(80-\sqrt[3]{14336+\sqrt{-5549064193}}+\sqrt[3]{14336-\sqrt{-5549064193}}\right)}$
15	$\frac{1}{8}\left(1+\sqrt{5}+\sqrt{30-6\sqrt{5}}\right)$
16	$\frac{1}{2}\left(\sqrt{2+\sqrt{2}}\right)$
17	$\frac{1}{16}\left(-1+\sqrt{17}+\sqrt{34-2\sqrt{17}}+2\sqrt{17+3\sqrt{17}-\sqrt{34-2\sqrt{17}}-2\sqrt{34+2\sqrt{17}}}\right)$
18	$\frac{i}{4}\left(\sqrt[3]{4+4\sqrt{-3}}-\sqrt[3]{4-4\sqrt{-3}}\right)$
19	
20	$\frac{1}{4}\left(\sqrt{10+2\sqrt{5}}\right)$
21	
22	
23	
24	$\frac{1}{4}\left(\sqrt{6}+\sqrt{2}\right)$

n	$\tan\left(\frac{2\pi}{n}\right)$
1	0
2	0

3	$-\sqrt{3}$
4	$\pm\infty$
5	$\sqrt{5+2\sqrt{5}}$
6	$\sqrt{3}$
7	
8	1
9	
10	$\sqrt{5-2\sqrt{5}}$
11	
12	$\frac{1}{3}\sqrt{3}$
13	
14	$\sqrt{\frac{112-\sqrt[3]{14336+\sqrt{-5549064193}}-\sqrt[3]{14336-\sqrt{-5549064193}}}{80+\sqrt[3]{14336+\sqrt{-5549064193}}+\sqrt[3]{14336-\sqrt{-5549064193}}}}$
15	$\frac{1}{2}\left(-3\sqrt{3}-\sqrt{15}+\sqrt{50+22\sqrt{5}}\right)$
16	$\sqrt{2}-1$
17	
18	
19	
20	$\frac{1}{5}\left(\sqrt{25-10\sqrt{5}}\right)$
21	
22	
23	
24	$2-\sqrt{3}$

Uses for Constants

As an example of the use of these constants, consider the volume of a regular dodecahedron, where a is the length of an edge:

$$V = \frac{5a^3 \cos 36^\circ}{\tan^2 36^\circ}.$$

Using,

$$\cos 36^\circ = \frac{\sqrt{5}+1}{4},$$

$$\tan 36^\circ = \sqrt{5-2\sqrt{5}},$$

this can be simplified to:

$$V = \frac{a^3\left(15+7\sqrt{5}\right)}{4}.$$

Derivation Triangles

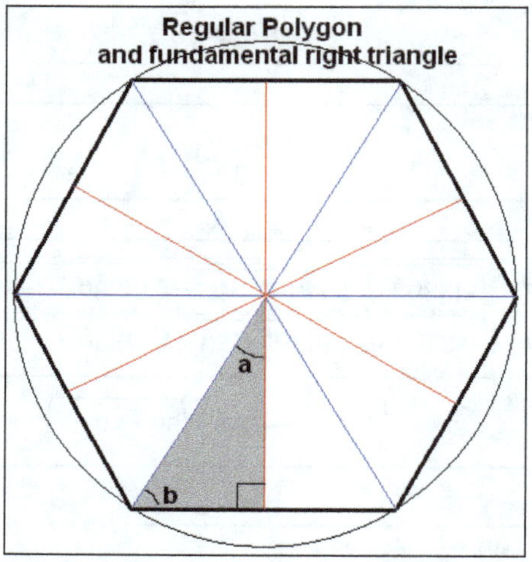

Regular polygon (n-sided) and its fundamental right triangle. Angles: $a = \frac{180^\circ}{n}$ and $b = 90\left(1-\frac{2}{n}\right)^\circ$..

The derivation of sine, cosine, and tangent constants into radial forms is based upon the constructibility of right triangles.

Here right triangles made from symmetry sections of regular polygons are used to calculate fundamental trigonometric ratios. Each right triangle represents three points in a regular polygon: a vertex, an edge center containing that vertex, and the polygon center. An n-gon can be divided into $2n$ right triangles with angles of $\frac{180}{n}$, $90 - \frac{180}{n}$, 90} degrees, for n in 3, 4, 5, ...

Constructibility of 3, 4, 5, and 15-sided polygons are the basis, and angle bisectors allow multiples of two to also be derived.

- Constructible:
  - 3×2^n-sided regular polygons, for n = 0, 1, 2, 3, ...
    - 30°-60°-90° triangle: Triangle (3-sided).
    - 60°-30°-90° triangle: Hexagon (6-sided).

- $75°$-$15°$-$90°$ triangle: Dodecagon (12-sided).
- $82.5°$-$7.5°$-$90°$ triangle: Icositetragon (24-sided).
- $86.25°$-$3.75°$-$90°$ triangle: Tetracontaoctagon (48-sided).
- $88.125°$-$1.875°$-$90°$ triangle: Enneacontahexagon (96-sided).
- $89.0625°$-$0.9375°$-$90°$ triangle: 192-gon.
- $89.53125°$-$0.46875°$-$90°$ triangle: 384-gon, etc.

- 4×2^n-sided:
  - $45°$-$45°$-$90°$ triangle: Square (4-sided).
  - $67.5°$-$22.5°$-$90°$ triangle: Octagon (8-sided).
  - $78.75°$-$11.25°$-$90°$ triangle: Hexadecagon (16-sided).
  - $84.375°$-$5.625°$-$90°$ triangle: Triacontadigon (32-sided).
  - $87.1875°$-$2.8125°$-$90°$ triangle: Hexacontatetragon (64-sided).
  - $88.09375°$-$1.40625°$-$90°$ triangle: 128-gon.
  - $89.046875°$-$0.703125°$-$90°$ triangle: 256-gon, etc.
- 5×2^n-sided:
  - $54°$-$36°$-$90°$ triangle: Pentagon (5-sided).
  - $72°$-$18°$-$90°$ triangle: Decagon (10-sided).
  - $81°$-$9°$-$90°$ triangle: Icosagon (20-sided).
  - $85.5°$-$4.5°$-$90°$ triangle: Tetracontagon (40-sided).
  - $87.75°$-$2.25°$-$90°$ triangle: Octacontagon (80-sided).
  - $88.875°$-$1.125°$-$90°$ triangle: 160-gon.
  - $89.4375°$-$0.5625°$-$90°$ triangle: 320-gon, etc.
- 15×2^n-sided:
  - $78°$-$12°$-$90°$ triangle: Pentadecagon (15-sided).
  - $84°$-$6°$-$90°$ triangle: Triacontagon (30-sided).
  - $87°$-$3°$-$90°$ triangle: Hexacontagon (60-sided).

    - 88.5°-1.5°-90° triangle: Hecatonicosagon (120-sided).
    - 89.25°-0.75°-90° triangle: 240-gon, etc.

  There are also higher constructible regular polygons: 17, 51, 85, 255, 257, 353, 449, 641, 1409, 2547, ..., 65535, 65537, 69481, 73697, ..., 4294967295).

- Nonconstructible (with whole or half degree angles): No finite radical expressions involving real numbers for these triangle edge ratios are possible, therefore its multiples of two are also not possible.
  - 9×2^n-sided:
    - 70°-20°-90° triangle: Enneagon (9-sided).
    - 80°-10°-90° triangle: Octadecagon (18-sided).
    - 85°-5°-90° triangle: Triacontahexagon (36-sided).
    - 87.5°-2.5°-90° triangle: Heptacontadigon (72-sided).
  - 45×2^n-sided:
    - 86°-4°-90° triangle: Tetracontapentagon (45-sided).
    - 88°-2°-90° triangle: Enneacontagon (90-sided).
    - 89°-1°-90° triangle: 180-gon.
    - 89.5°-0.5°-90° triangle: 360-gon.

Calculated Trigonometric Values for Sine and Cosine

The Trivial Values

In degree format, sin and cos of 0, 30, 45, 60, and 90 can be calculated from their right angled triangles, using the Pythagorean theorem.

In radian format, sin and cos of $\pi/2^n$ can be expressed in radical format by recursively applying the following:

$$2\cos\theta = \sqrt{2+2\cos 2\theta} = \sqrt{2+\sqrt{2+2\cos 4\theta}} = \sqrt{2+\sqrt{2+\sqrt{2+2\cos 8\theta}}}$$

and so on.

$$2\sin\theta = \sqrt{2-2\cos 2\theta} = \sqrt{2-\sqrt{2+2\cos 4\theta}} = \sqrt{2-\sqrt{2+\sqrt{2+2\cos 8\theta}}}$$

and so on.

For example:

$$\cos\frac{\pi}{2^1}=\frac{0}{2}$$

$$\cos\frac{\pi}{2^2}=\frac{\sqrt{2+0}}{2} \text{ and } \sin\frac{\pi}{2^2}=\frac{\sqrt{2-0}}{2}$$

$$\cos\frac{\pi}{2^3}=\frac{\sqrt{2+\sqrt{2}}}{2} \text{ and } \sin\frac{\pi}{2^3}=\frac{\sqrt{2-\sqrt{2}}}{2}$$

$$\cos\frac{\pi}{2^4}=\frac{\sqrt{2+\sqrt{2+\sqrt{2}}}}{2} \text{ and } \sin\frac{\pi}{2^4}=\frac{\sqrt{2-\sqrt{2+\sqrt{2}}}}{2}$$

$$\cos\frac{\pi}{2^5}=\frac{\sqrt{2+\sqrt{2+\sqrt{2+\sqrt{2}}}}}{2} \text{ and } \sin\frac{\pi}{2^5}=\frac{\sqrt{2-\sqrt{2+\sqrt{2+\sqrt{2}}}}}{2}$$

$$\cos\frac{\pi}{2^6}=\frac{\sqrt{2+\sqrt{2+\sqrt{2+\sqrt{2+\sqrt{2}}}}}}{2} \text{ and } \sin\frac{\pi}{2^6}=\frac{\sqrt{2-\sqrt{2+\sqrt{2+\sqrt{2+\sqrt{2}}}}}}{2}$$

and so on.

Radical form, Sin and Cos of $\frac{\pi}{(3\times 2^n)}$

$$\cos\frac{2\pi}{3}=\frac{-1}{2}$$

$$\cos\frac{\pi}{3\times 2^0}=\frac{\sqrt{2-1}}{2} \text{ and } \sin\frac{\pi}{3\times 2^0}=\frac{\sqrt{2+1}}{2}$$

$$\cos\frac{\pi}{3\times 2^1}=\frac{\sqrt{2+1}}{2} \text{ and } \sin\frac{\pi}{3\times 2^1}=\frac{\sqrt{2-1}}{2}$$

$$\cos\frac{\pi}{3\times 2^2}=\frac{\sqrt{2+\sqrt{3}}}{2} \text{ and } \sin\frac{}{3\times 2} \quad \overline{\sqrt{2\quad\sqrt{3}}}$$

$$\cos\frac{\pi}{3\times 2^3}=\frac{\sqrt{2+\sqrt{2+\sqrt{3}}}}{2} \text{ and } \sin\frac{\pi}{3\times 2^3}=\frac{\sqrt{2-\sqrt{2+\sqrt{3}}}}{2}$$

$$\cos\frac{\pi}{3\times 2^4}=\frac{\sqrt{2+\sqrt{2+\sqrt{2+\sqrt{3}}}}}{2} \text{ and } \sin\frac{\pi}{3\times 2^4}=\frac{\sqrt{2-\sqrt{2+\sqrt{2+\sqrt{3}}}}}{2}$$

$$\cos\frac{\pi}{3\times 2^5}=\frac{\sqrt{2+\sqrt{2+\sqrt{2+\sqrt{2+\sqrt{3}}}}}}{2}\text{ and }\sin\frac{\pi}{3\times 2^5}=\frac{\sqrt{2-\sqrt{2+\sqrt{2+\sqrt{2+\sqrt{3}}}}}}{2}$$

and so on.

Radical form, Sin and Cos of $\frac{\pi}{(5\times 2^n)}$

$$\cos\frac{2\pi}{5}=\frac{\sqrt{5}-1}{4}$$

$$\cos\frac{\pi}{5\times 2^0}=\frac{\sqrt{5}+1}{4}\text{ (Therefore }2+2\cos\frac{\pi}{5}=2+\sqrt{1.25}+0.5\text{)}$$

$$\cos\frac{\pi}{5\times 2^1}=\frac{\sqrt{2.5+\sqrt{1.25}}}{2}\text{ and }\sin\frac{\pi}{5\times 2^1}=\frac{\sqrt{1.5-\sqrt{1.25}}}{2}$$

$$\cos\frac{}{5\times 2}\quad\frac{\sqrt{2+\sqrt{2.5+\sqrt{1.25}}}}{}\text{ and }\sin\frac{\pi}{5\times 2^2}=\frac{\sqrt{2-\sqrt{2.5+\sqrt{1.25}}}}{2}$$

$$\cos\frac{\pi}{5\times 2^3}=\frac{\sqrt{2+\sqrt{2+\sqrt{2.5+\sqrt{1.25}}}}}{2}\text{ and}$$

$$\sin\frac{\pi}{5\times 2^3}=\frac{\sqrt{2-\sqrt{2+\sqrt{2.5+\sqrt{1.25}}}}}{2}$$

$$\cos\frac{\pi}{5\times 2^4}=\frac{\sqrt{2+\sqrt{2+\sqrt{2+\sqrt{2.5+\sqrt{1.25}}}}}}{2}\text{ and}$$

$$\sin\frac{\pi}{5\times 2^4}=\frac{\sqrt{2-\sqrt{2+\sqrt{2+\sqrt{2.5+\sqrt{1.25}}}}}}{2}$$

$$\cos\frac{\pi}{5\times 2^5}=\frac{\sqrt{2+\sqrt{2+\sqrt{2+\sqrt{2+\sqrt{2.5+\sqrt{1.25}}}}}}}{2}\text{ and}$$

$$\sin\frac{\pi}{5\times 2^5}=\frac{\sqrt{2-\sqrt{2+\sqrt{2+\sqrt{2+\sqrt{2.5+\sqrt{1.25}}}}}}}{2}$$

and so on.

Radical form, Sin and Cos of $\frac{\pi}{(5\times3\times2^{n})}$

$$\cos\frac{\pi}{15\times2^{0}}=\frac{\sqrt{\sqrt{0.703125}+1.875+\sqrt{0.3125}-0.25}}{2}$$

$$\cos\frac{\pi}{15\times2^{1}}=\frac{\sqrt{\sqrt{\sqrt{0.703125}+1.875}+\sqrt{0.3125}+1.75}}{2} \text{ and}$$

$$\sin\frac{\pi}{15\times2^{1}}=\frac{\sqrt{2.25-\sqrt{\sqrt{0.703125}+1.875}-\sqrt{0.3125}}}{2}$$

$$\cos\frac{\pi}{15\times2^{2}}=\frac{\sqrt{2+\sqrt{\sqrt{\sqrt{0.703125}+1.875}+\sqrt{0.3125}+1.75}}}{2} \text{ and}$$

$$\sin\frac{\pi}{15\times2^{2}}=\frac{\sqrt{2-\sqrt{\sqrt{\sqrt{0.703125}+1.875}+\sqrt{0.3125}+1.75}}}{2}$$

$$\cos\frac{\pi}{15\times2^{3}}=\frac{\sqrt{2+\sqrt{2+\sqrt{\sqrt{\sqrt{0.703125}+1.875}+\sqrt{0.3125}+1.75}}}}{2} \text{ and}$$

$$\sin\frac{\pi}{15\times2^{3}}=\frac{\sqrt{2-\sqrt{2+\sqrt{\sqrt{\sqrt{0.703125}+1.875}+\sqrt{0.3125}+1.75}}}}{2}$$

$$\cos\frac{\pi}{15\times2^{4}}=\frac{\sqrt{2+\sqrt{2+\sqrt{2+\sqrt{\sqrt{\sqrt{0.703125}+1.875}+\sqrt{0.3125}+1.75}}}}}{2} \text{ and}$$

$$\sin\frac{\pi}{15\times2^{4}}=\frac{\sqrt{2-\sqrt{2+\sqrt{2+\sqrt{\sqrt{\sqrt{0.703125}+1.875}+\sqrt{0.3125}+1.75}}}}}{2}$$

$$\cos\frac{\pi}{15\times2^{5}}=\frac{\sqrt{2+\sqrt{2+\sqrt{2+\sqrt{2+\sqrt{\sqrt{\sqrt{0.703125}+1.875}+\sqrt{0.3125}+1.75}}}}}}{2} \text{ and}$$

$$\sin\frac{\pi}{15\times2^{5}}=\frac{\sqrt{2-\sqrt{2+\sqrt{2+\sqrt{2+\sqrt{\sqrt{\sqrt{0.703125}+1.875}+\sqrt{0.3125}+1.75}}}}}}{2}$$

and so on.

Radical form, Sin and Cos of $\frac{\pi}{\left(17\times 2^{n}\right)}$

If $M = 2(17+\sqrt{17})$ and $N = 2(17-\sqrt{17})$ then

$$\cos\frac{\pi}{17} = \frac{\sqrt{M-4+2(\sqrt{N}+\sqrt{2(2M-N+\sqrt{17N}-\sqrt{N}-8\sqrt{M}))}}}{8}.$$

Therefore, applying induction:

$$\cos\frac{\pi}{17\times 2^{0}} = \frac{\sqrt{30+2\sqrt{17}+\sqrt{136-8\sqrt{17}}+\sqrt{272+48\sqrt{17}+8\sqrt{34-2\sqrt{17}}\times(\sqrt{17}-1)-64\sqrt{34+2\sqrt{17}}}}}{8},$$

$$\cos\frac{\pi}{17\times 2^{n+1}} = \frac{\sqrt{2+2\cos\frac{\pi}{17\times 2^{n}}}}{2} \text{ and } \sin\frac{\pi}{17\times 2^{n+1}} = \frac{\sqrt{2-2\cos\frac{\pi}{17\times 2^{n}}}}{2}.$$

Radical form, Sin and Cos of $\frac{\pi}{\left(257\times 2^{n}\right)}$ and $\frac{\pi}{\left(65537\times 2^{n}\right)}$

The induction above can be applied in the same way to all the remaining Fermat primes ($F_3 = 2^{2^3}+1 = 2^8+1 = 257$ and $F_4 = 2^{2^4}+1 = 2^{16}+1 = 65537$), the factors of π whose cos and sin radical expressions are known to exist but are very long to express here.

$$\cos\frac{\pi}{257\times 2^{n+1}} = \frac{\sqrt{2+2\cos\frac{\pi}{257\times 2^{n}}}}{2} \text{ and } \sin\frac{\pi}{257\times 2^{n+1}} = \frac{\sqrt{2-2\cos\frac{\pi}{257\times 2^{n}}}}{2};$$

$$\cos\frac{\pi}{65537\times 2^{n+1}} = \frac{\sqrt{2+2\cos\frac{\pi}{65537\times 2^{n}}}}{2} \text{ and } \sin\frac{}{65537\times 2^{n\ 1}} \quad \frac{\sqrt{2\quad 2\cos\frac{}{65537\times 2}}}{}.$$

Radical form, Sin and Cos of $\frac{\pi}{\left(255\times 2^{n}\right)}$, $\frac{\pi}{\left(65535\times 2^{n}\right)}$ and $\frac{\pi}{\left(4294967295\times 2^{n}\right)}$

$D = 2^{32} - 1 = 4{,}294{,}967{,}295$ is the largest odd integer denominator for which radical forms for $\sin(\pi/D)$ and $\cos(\pi/D)$ are known to exist.

Using the radical form values from the sections above, and applying cos(A-B) = cosA cosB + sinA sinB, followed by induction, we get:

$$\cos\frac{\pi}{255\times 2^0}=\frac{\sqrt{2+2\cos\left(\frac{\pi}{15}-\frac{\pi}{17}\right)}}{2} \text{ and } \sin\frac{\pi}{255\times 2^0}=\frac{\sqrt{2-2\cos\frac{\pi}{15}-\frac{\pi}{17}}}{2};$$

$$\cos\frac{\pi}{255\times 2^{n+1}}=\frac{\sqrt{2+2\cos\frac{\pi}{255\times 2^n}}}{2} \text{ and } \sin\frac{\pi}{255\times 2^{n+1}}=\frac{\sqrt{2-2\cos\frac{\pi}{255\times 2^n}}}{2}.$$

Therefore, using the radical form values from the sections above, and applying cos(A-B) = cosA cosB + sinA sinB, followed by induction, we get:

$$\cos\frac{\pi}{65535\times 2^0}=\frac{\sqrt{2+2\cos\left(\frac{\pi}{255}-\frac{\pi}{257}\right)}}{2} \text{ and}$$

$$\sin\frac{\pi}{65535\times 2^0}=\frac{\sqrt{2-2\cos\left(\frac{\pi}{255}-\frac{\pi}{257}\right)}}{2};$$

$$\cos\frac{\pi}{65535\times 2^{n+1}}=\frac{\sqrt{2+2\cos\frac{\pi}{65535\times 2^n}}}{2} \text{ and}$$

$$\sin\frac{\pi}{65535\times 2^{n+1}}=\frac{\sqrt{2-2\cos\frac{\pi}{65535\times 2^n}}}{2}.$$

Finally, using the radical form values from the sections above, and applying cos(A-B) = cosA cosB + sinA sinB, followed by induction, we get:

$$\cos\frac{\pi}{4294967295\times 2^0}=\frac{\sqrt{2+2\cos\left(\frac{\pi}{65535}-\frac{\pi}{65537}\right)}}{2} \text{ and}$$

$$\sin\frac{\pi}{4294967295\times 2^0}=\frac{\sqrt{2-2\cos(\frac{\pi}{65535}-\frac{\pi}{65537})}}{2};$$

$$\cos\frac{\pi}{4294967295\times 2^{n+1}}=\frac{\sqrt{2+2\cos\frac{\pi}{4294967295\times 2^n}}}{2} \text{ and}$$

$$\sin\frac{\pi}{4294967295\times 2^{n+1}}=\frac{\sqrt{2-2\cos\frac{\pi}{4294967295\times 2^{n}}}}{2}.$$

The radical form expansion of the above is very large, hence expressed in the simpler form above.

$$\mathbf{n}\times\frac{\pi}{\left(5\times 2^{m}\right)}$$

Geometrical Method

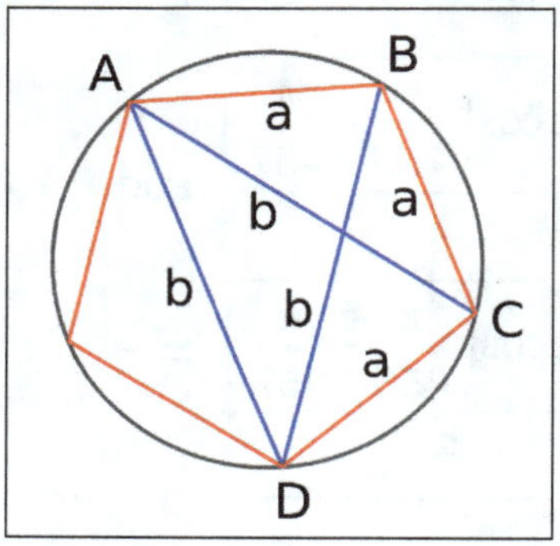

Chord $(36°)=\frac{a}{b}=\frac{1}{\varphi}$, i.e., the reciprocal of the golden ratio, from Ptolemy's theorem.

Applying Ptolemy's theorem to the cyclic quadrilateral ABCD defined by four successive vertices of the pentagon, we can find that:

$$\mathrm{crd}\,36^\circ=\mathrm{crd}(\angle ADB)=\frac{a}{b}=\frac{2}{1+\sqrt{5}}=\frac{\sqrt{5}-1}{2}$$

which is the reciprocal $\frac{1}{\varphi}$ of the golden ratio. crd is the chord function,

$$\mathrm{crd}\,\theta=2\sin\frac{\theta}{2}.$$

Thus

$$\sin 18^\circ=\frac{1}{1+\sqrt{5}}=\frac{\sqrt{5}-1}{4}.$$

(Alternatively, without using Ptolemy's theorem, label as X the intersection of AC and BD, and note by considering angles that triangle AXB is isosceles, so AX = AB = *a*.

Triangles AXD and CXB are similar, because AD is parallel to BC. So $XC = a\cdot\left(\frac{a}{b}\right)$. But AX + XC = AC, so $a+\frac{a^2}{b}=b$. Solving this gives $\frac{a}{b}=\frac{1}{\varphi}$, as above).

Similarly,

$$\operatorname{crd} 108^\circ = \operatorname{crd}(\angle ABC) = \frac{b}{a} = \frac{1+\sqrt{5}}{2},$$

so

$$\sin 54^\circ = \cos 36^\circ = \frac{1+\sqrt{5}}{4}.$$

Algebraic Method

If θ is 18° or -54°, then 2θ and 3θ add up to 5θ = 90° or -270°, therefore sin 2θ is equal to cos 3θ.

$$(2\sin\theta)\cos\theta = \sin 2\theta = \cos 3\theta = 4\cos^3\theta - 3\cos\theta = (4\cos^2\theta - 3)\cos\theta$$
$$= (1-4\sin^2\theta)\cos\theta$$

So, $4\sin^2\theta + 2\sin\theta - 1 = 0$, which implies $\sin\theta = \sin(18^\circ, -54^\circ) = \frac{-1\pm\sqrt{5}}{4}$.

Therefore,

$$\sin(18^\circ) = \cos(72^\circ) = \frac{\sqrt{5}-1}{4} \text{ and } \sin(54^\circ) = \cos(36^\circ) = \frac{\sqrt{5}+1}{4} \text{ and}$$

$$\sin(36^\circ) = \cos(54^\circ) = \frac{\sqrt{10-2\sqrt{5}}}{4} \text{ and } \sin(72^\circ) = \cos(18^\circ) = \frac{\sqrt{10+2\sqrt{5}}}{4}.$$

Alternately, the multiple-angle formulas for functions of $5x$, where $x \in \{18, 36, 54, 72, 90\}$ and $5x \in \{90, 180, 270, 360, 450\}$, can be solved for the functions of x, since we know the function values of $5x$. The multiple-angle formulas are:

$$\sin 5x = 16\sin^5 x - 20\sin^3 x + 5\sin x,$$

$$\cos 5x = 16\cos^5 x - 20\cos^3 x + 5\cos x.$$

- When $\sin 5x = 0$ or $\cos 5x = 0$, we let $y = \sin x$ or $y = \cos x$ and solve for y:

  $$16y^5 - 20y^3 + 5y = 0.$$

  One solution is zero, and the resulting quartic equation can be solved as a quadratic in y^2.

- When sin $5x = 1$ or cos $5x = 1$, we again let $y = \sin x$ or $y = \cos x$ and solve for y:

  $16y^5 - 20y^3 + 5y - 1 = 0,$

  which factors into:

  $(y-1)(4y^2 + 2y - 1)^2 = 0.$

  $\mathbf{n \times \frac{\pi}{20}}$

  9° is 45 – 36, and 27° is 45 – 18; so we use the subtraction formulas for sine and cosine.

  $\mathbf{n \times \frac{\pi}{30}}$

  6° is 36 – 30, 12° is 30 – 18, 24° is 54 – 30, and 42° is 60 – 18; so we use the subtraction formulas for sine and cosine.

  $\mathbf{n \times \frac{\pi}{60}}$

  3° is 18 – 15, 21° is 36 – 15, 33° is 18 + 15, and 39° is 54 – 15, so we use the subtraction (or addition) formulas for sine and cosine.

Strategies for Simplifying Expressions

- Rationalizing the denominator: If the denominator is a square root, multiply the numerator and denominator by that radical. If the denominator is the sum or difference of two terms, multiply the numerator and denominator by the conjugate of the denominator. The conjugate is the identical, except the sign between the terms is changed. Sometimes you need to rationalize the denominator more than once.

- Splitting a fraction in two: Sometimes it helps to split the fraction into the sum of two fractions and then simplify both separately.

- Squaring and taking square roots: If there is a complicated term, with only one kind of radical in a term, this plan may help. Square the term, combine like terms, and take the square root. This may leave a big radical with a smaller radical inside, but it is often better than the original.

- Simplifying nested radical expressions: In general nested radicals cannot be reduced. But if,

  $\sqrt{a \pm b\sqrt{c}}$ with a, b, and c rational, we have $R = \sqrt{a^2 - b^2c}$

is rational, then both

$$d = \frac{a+R}{2} \text{ and } e = \frac{a-R}{2}$$

are rational; then we have

$$\sqrt{a \pm b\sqrt{c}} = \sqrt{d} \pm \sqrt{e}.$$

For example,

$$4\sin 18^\circ = \sqrt{6-2\sqrt{5}} = \sqrt{5}-1.$$

$$4\sin 15^\circ = 2\sqrt{2-\sqrt{3}} = \sqrt{2}\left(\sqrt{3}-1\right).$$

HERMITE'S COTANGENT IDENTITY

In mathematics, Hermite's cotangent identity is a trigonometric identity discovered by Charles Hermite. Suppose $a_1, \ldots, a_n$ are complex numbers, no two of which differ by an integer multiple of π. Let

$$A_{n,k} = \prod_{\substack{1 \le j \le n \\ j \ne k}} \cot(a_k - a_j)$$

(in particular, $A_{1,1}$, being an empty product, is 1). Then,

$$\cot(z-a_1)\cdots\cot(z-a_n) = \cos\frac{n\pi}{2} + \sum_{k=1}^{n} A_{n,k}\cot(z-a_k).$$

The simplest non-trivial example is the case $n = 2$:

$$\cot(z-a_1)\cot(z-a_2) = -1 + \cot(a_1-a_2)\cot(z-a_1) + \cot(a_2-a_1)\cot(z-a_2).$$

MORRIE'S LAW

Morrie's law is the trigonometric identity,

$$\cos(20^\circ)\cdot\cos(40^\circ)\cdot\cos(80^\circ) = \frac{1}{8}.$$

It is a special case of the more general identity,

$$2^n \cdot \prod_{k=0}^{n-1}\cos(2^k\alpha) = \frac{\sin(2^n\alpha)}{\sin(\alpha)}$$

with $n = 3$ and $\alpha = 20^\circ$ and the fact that,

$$\frac{\sin(160^\circ)}{\sin(20^\circ)} = \frac{\sin(180^\circ - 20^\circ)}{\sin(20^\circ)} = 1,$$

since,

$$\sin(180^\circ - x) = \sin(x).$$

A similar identity for the sine function also holds:

$$\sin(20^\circ)\cdot\sin(40^\circ)\cdot\sin(80^\circ) = \frac{\sqrt{3}}{8}.$$

Moreover, dividing the second identity by the first, the following identity is evident:

$$\tan(20^\circ)\cdot\tan(40^\circ)\cdot\tan(80^\circ) = \sqrt{3} = \tan(60^\circ).$$

Proof:

Recall the double angle formula for the sine function,

$$\sin(2\alpha) = 2\sin(\alpha)\cos(\alpha).$$

Solve for $\cos(\alpha)$

$$\cos(\alpha) = \frac{\sin(2\alpha)}{2\sin(\alpha)}.$$

It follows that:

$$\cos(2\alpha) = \frac{\sin(4\alpha)}{2\sin(2\alpha)}$$

$$\cos(4\alpha) = \frac{\sin(8\alpha)}{2\sin(4\alpha)}$$

$$\vdots$$

$$\cos(2^{n-1}\alpha) = \frac{\sin(2^n\alpha)}{2\sin(2^{n-1}\alpha)}.$$

Multiplying all of these expressions together yields:

$$\cos(\alpha)\cos(2\alpha)\cos(4\alpha)\cdots\cos(2^{n-1}\alpha) = \frac{\sin(2\alpha)}{2\sin(\alpha)}\cdot\frac{\sin(4\alpha)}{2\sin(2\alpha)}\cdot\frac{\sin(8\alpha)}{2\sin(4\alpha)}\cdots\frac{\sin(2^n\alpha)}{2\sin(2^{n-1}\alpha)}.$$

The intermediate numerators and denominators cancel leaving only the first denominator, a power of 2 and the final numerator. Note that there are n terms in both sides of the expression.

Thus,

$$\prod_{k=0}^{n-1} \cos(2^k \alpha) = \frac{\sin(2^n \alpha)}{2^n \sin(\alpha)},$$

which is equivalent to the generalization of Morrie's law.

PYTHAGOREAN TRIGONOMETRIC IDENTITY

The Pythagorean trigonometric identity, also called the fundamental Pythagorean trigonometric identity or simply Pythagorean identity is an identity expressing the Pythagorean theorem in terms of trigonometric functions. Along with the sum-of-angles formulae, it is one of the basic relations between the sine and cosine functions.

The identity is,

$$\sin^2 \theta + \cos^2 \theta = 1.$$

As usual, $\sin^2 \theta$ means $(\sin \theta)^2$.

Proofs and their Relationships to the Pythagorean Theorem

Proof Based on Right-angle Triangles

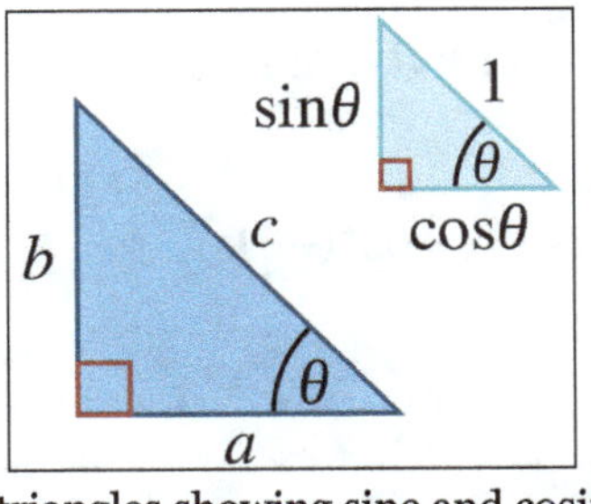

Similar right triangles showing sine and cosine of angle θ.

Any similar triangles have the property that if we select the same angle in all of them, the ratio of the two sides defining the angle is the same regardless of which similar triangle is selected, regardless of its actual size: the ratios depend upon the three angles, not the lengths of the sides. Thus for either of the similar right triangles in the figure, the ratio of its horizontal side to its hypotenuse is the same, namely $\cos \theta$.

The elementary definitions of the sine and cosine functions in terms of the sides of a right triangle are:

$$\sin \theta = \frac{\text{opposite}}{\text{hypotenuse}} = \frac{b}{c}$$

$$\cos\theta = \frac{\text{adjacent}}{\text{hypotenuse}} = \frac{a}{c}$$

The Pythagorean identity follows by squaring both definitions above, and adding; the left-hand side of the identity then becomes,

$$\frac{\text{opposite}^2 + \text{adjacent}^2}{\text{hypotenuse}^2}$$

which by the Pythagorean theorem is equal to 1. This definition is valid for all angles, due to the definition of defining $x = \cos\theta$ and $y = \sin\theta$ for the unit circle and thus $x = c\cos\theta$ and $y = c\sin\theta$ for a circle of radius c and reflecting our triangle in the y axis and setting $a = x$ and $b = y$.

Alternatively, the identities found at Trigonometric symmetry, shifts, and periodicity may be employed. By the periodicity identities we can say if the formula is true for $-\pi < \theta \le \pi$ then it is true for all real θ. Next we prove the range $\pi/2 < \theta \le \pi$, to do this we let $t = \theta - \pi/2$, t will now be in the range $0 < t \le \pi/2$. We can then make use of squared versions of some basic shift identities (squaring conveniently removes the minus signs):

$$\sin^2\theta + \cos^2\theta = \sin^2\left(t + \frac{1}{2}\pi\right) + \cos^2\left(t + \frac{1}{2}\pi\right) = \cos^2 t + \sin^2 t = 1.$$

All that remains is to prove it for $-\pi < \theta < 0$; this can be done by squaring the symmetry identities to get,

$$\sin^2\theta = \sin^2(-\theta) \text{ and } \cos^2\theta = \cos^2(-\theta).$$

Related Identities

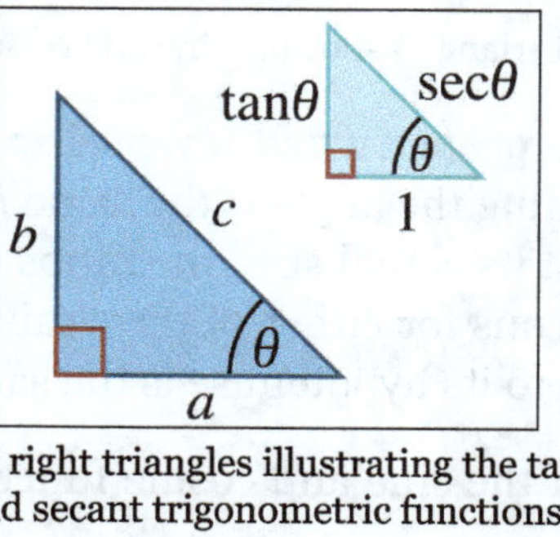

Similar right triangles illustrating the tangent and secant trigonometric functions.

The identities,

$$1 + \tan^2\theta = \sec^2\theta$$

and

$$1+\cot^2\theta=\csc^2\theta$$

are also called Pythagorean trigonometric identities. If one leg of a right triangle has length 1, then the tangent of the angle adjacent to that leg is the length of the other leg, and the secant of the angle is the length of the hypotenuse.

$$\tan\theta=\frac{b}{a},$$

and

$$\sec\theta=\frac{c}{a}.$$

In this way, this trigonometric identity involving the tangent and the secant follows from the Pythagorean theorem. The angle opposite the leg of length 1 (this angle can be labeled $\varphi = \pi/2 - \theta$) has cotangent equal to the length of the other leg, and cosecant equal to the length of the hypotenuse. In that way, this trigonometric identity involving the cotangent and the cosecant also follows from the Pythagorean theorem.

The following table gives the identities with the factor or divisor that relates them to the main identity:

Original Identity	Divisor	Divisor Equation	Derived Identity	Derived Identity (Alternate)
$\sin^2\theta+\cos^2\theta=1$	$\cos^2\theta$	$\frac{\sin^2\theta}{\cos^2\theta}+\frac{\cos^2\theta}{\cos^2\theta}=\frac{1}{\cos^2\theta}$	$\tan^2\theta+1=\sec^2\theta$	$\tan^2\theta=\sec^2\theta-1$
$\sin^2\theta+\cos^2\theta=1$	$\sin^2\theta$	$\frac{\sin^2\theta}{\sin^2\theta}+\frac{\cos^2\theta}{\sin^2\theta}=\frac{1}{\sin^2\theta}$	$1+\cot^2\theta=\csc^2\theta$	$\cot^2\theta=\csc^2\theta-1$

Proof using the Unit Circle

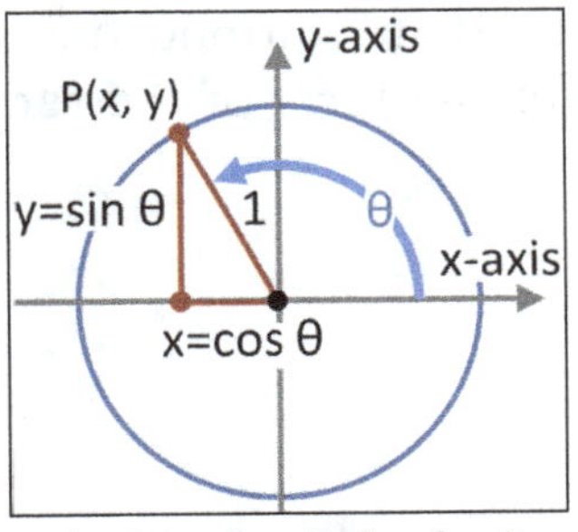

Point P(x,y) on the circle of unit radius at an obtuse angle $\theta > \pi/2$.

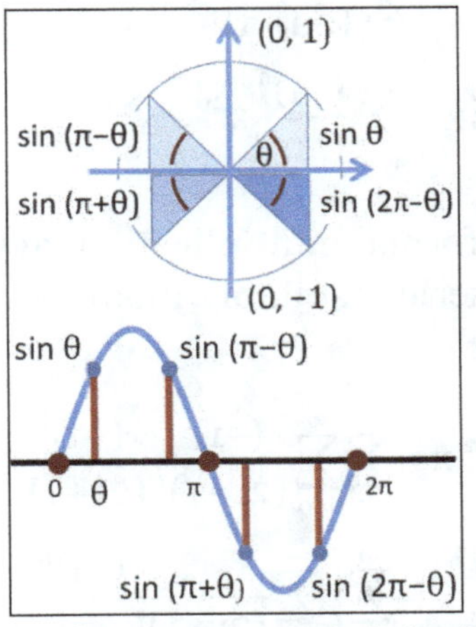

Sine function on unit circle (top) and its graph (bottom).

The unit circle centered at the origin in the Euclidean plane is defined by the equation:

$$x^2 + y^2 = 1.$$

Given an angle θ, there is a unique point *P* on the unit circle at an angle θ from the *x*-axis, and the *x*- and *y*-coordinates of *P* are:

$$x = \cos\theta \text{ and } y = \sin\theta.$$

Consequently, from the equation for the unit circle:

$$\cos^2\theta + \sin^2\theta = 1,$$

the Pythagorean identity.

In the figure, the point P has a negative x-coordinate, and is appropriately given by x = cosθ, which is a negative number: cosθ = −cos(π−θ). Point P has a positive y-coordinate, and sinθ = sin(π−θ) > 0. As θ increases from zero to the full circle θ = 2π, the sine and cosine change signs in the various quadrants to keep x and y with the correct signs. The figure shows how the sign of the sine function varies as the angle changes quadrant.

Because the *x*- and *y*-axes are perpendicular, this Pythagorean identity is equivalent to the Pythagorean theorem for triangles with hypotenuse of length 1 (which is in turn equivalent to the full Pythagorean theorem by applying a similar-triangles argument).

Proof using Power Series

The trigonometric functions may also be defined using power series, namely (for *x* an angle measured in radians):

$$\sin x = \sum_{n=0}^{\infty} \frac{(-1)^n}{(2n+1)!} x^{2n+1},$$

$$\cos x = \sum_{n=0}^{\infty} \frac{(-1)^n}{(2n)!} x^{2n}.$$

Using the formal multiplication law for power series at Multiplication and division of power series (suitably modified to account for the form of the series here) we obtain,

$$\sin^2 x = \sum_{i=0}^{\infty}\sum_{j=0}^{\infty} \frac{(-1)^i}{(2i+1)!} \frac{(-1)^j}{(2j+1)!} x^{(2i+1)+(2j+1)}$$

$$= \sum_{n=1}^{\infty} \left(\sum_{i=0}^{n-1} \frac{(-1)^{n-1}}{(2i+1)!(2(n-i-1)+1)!} \right) x^{2n}$$

$$=\sum_{n=1}^{\infty}\left(\sum_{i=0}^{n-1}\binom{2n}{2i+1}\right)\frac{(-1)^{n-1}}{(2n)!}x^{2n},$$

$$\cos^2 x=\sum_{i=0}^{\infty}\sum_{j=0}^{\infty}\frac{(-1)^i}{(2i)!}\frac{(-1)^j}{(2j)!}x^{(2i)+(2j)}$$

$$=\sum_{n=0}^{\infty}\left(\sum_{i=0}^{n}\frac{(-1)^n}{(2i)!(2(n-i))!}\right)x^{2n}$$

$$=\sum_{n=0}^{\infty}\left(\sum_{i=0}^{n}\binom{2n}{2i}\right)\frac{(-1)^n}{(2n)!}x^{2n}.$$

In the expression for $\sin^2$, n must be at least 1, while in the expression for $\cos^2$, the constant term is equal to 1. The remaining terms of their sum are (with common factors removed).

$$\sum_{i=0}^{n}\binom{2n}{2i}-\sum_{i=0}^{n-1}\binom{2n}{2i+1}=\sum_{j=0}^{2n}(-1)^j\binom{2n}{j}=(1-1)^{2n}=0$$

by the binomial theorem. Consequently,

$$\sin^2 x+\cos^2 x=1,$$

which is the Pythagorean trigonometric identity.

The Pythagorean theorem is not closely related to the Pythagorean identity when the trigonometric functions are defined in this way; instead, in combination with the theorem, the identity now shows that these power series parameterize the unit circle.

Proof using the Differential Equation

Sine and cosine can be defined as the two solutions to the differential equation:

$$y''+y=0$$

satisfying respectively $y(0) = 0$, $y'(0) = 1$ and $y(0) = 1$, $y'(0) = 0$. It follows from the theory of ordinary differential equations that the first solution, sine, has the second, cosine, as its derivative, and it follows from this that the derivative of cosine is the negative of the sine. The identity is equivalent to the assertion that the function,

$$z=\sin^2 x+\cos^2 x$$

is constant and equal to 1. Differentiating using the chain rule gives:

$$\frac{d}{dx}z=2\sin x\cos x+2\cos x(-\sin x)=0,$$

so z is constant by the mean value theorem. A calculation confirms that $z(0) = 1$, and z is a constant so $z = 1$ for all x, so the Pythagorean identity is established.

A similar proof can be completed using power series as above to establish that the sine has as its derivative the cosine, and the cosine has as its derivative the negative sine. In fact, the definitions by ordinary differential equation and by power series lead to similar derivations of most identities.

This proof of the identity has no direct connection with Euclid's demonstration of the Pythagorean theorem.

PROOFS OF TRIGONOMETRIC IDENTITIES

The main trigonometric identities between trigonometric functions are proved, using mainly the geometry of the right triangle.

Elementary Trigonometric Identities

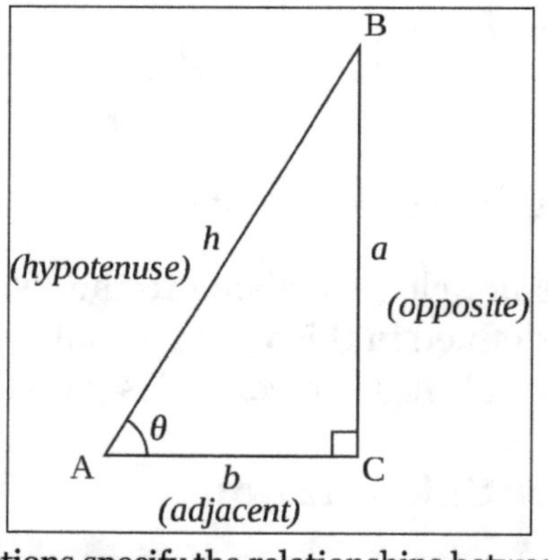

Trigonometric functions specify the relationships between side lengths and interior angles of a right triangle. For example, the sine of angle θ is defined as being the length of the opposite side divided by the length of the hypotenuse.

The six trigonometric functions are defined for every real number, except, for some of them, for angles that differ from 0 by a multiple of the right angle (90°). Referring to the diagram at the right, the six trigonometric functions of θ are, for angles smaller than the right angle:

$$\sin\theta = \frac{\text{opposite}}{\text{hypotenuse}} = \frac{a}{h}$$

$$\cos\theta = \frac{\text{adjacent}}{\text{hypotenuse}} = \frac{b}{h}$$

$$\tan\theta = \frac{\text{opposite}}{\text{adjacent}} = \frac{a}{b}$$

$$\cot\theta = \frac{\text{adjacent}}{\text{opposite}} = \frac{b}{a}$$

$$\sec\theta = \frac{\text{hypotenuse}}{\text{adjacent}} = \frac{h}{b}$$

$$\csc\theta = \frac{\text{hypotenuse}}{\text{opposite}} = \frac{h}{a}$$

Ratio Identities

In the case of angles smaller than a right angle, the following identities are direct consequences of above definitions through the division identity,

$$\frac{a}{b} = \frac{\left(\frac{a}{h}\right)}{\left(\frac{b}{h}\right)}.$$

They remain valid for angles greater than 90° and for negative angles.

$$\tan\theta = \frac{\text{opposite}}{\text{adjacent}} = \frac{\left(\frac{\text{opposite}}{\text{hypotenuse}}\right)}{\left(\frac{\text{adjacent}}{\text{hypotenuse}}\right)} = \frac{\sin\theta}{\cos\theta}$$

$$\cot\theta = \frac{\text{adjacent}}{\text{opposite}} = \frac{\left(\frac{\text{adjacent}}{\text{adjacent}}\right)}{\left(\frac{\text{opposite}}{\text{adjacent}}\right)} = \frac{1}{\tan\theta} = \frac{\cos\theta}{\sin\theta}$$

$$\sec\theta = \frac{1}{\cos\theta} = \frac{\text{hypotenuse}}{\text{adjacent}}$$

$$\csc\theta = \frac{1}{\sin\theta} = \frac{\text{hypotenuse}}{\text{opposite}}$$

$$\tan\theta = \frac{\text{opposite}}{\text{adjacent}} = \frac{\left(\frac{\text{opposite}\times\text{hypotenuse}}{\text{opposite}\times\text{adjacent}}\right)}{\left(\frac{\text{adjacent}\times\text{hypotenuse}}{\text{opposite}\times\text{adjacent}}\right)} = \frac{\left(\frac{\text{hypotenuse}}{\text{adjacent}}\right)}{\left(\frac{\text{hypotenuse}}{\text{opposite}}\right)} = \frac{\sec\theta}{\csc\theta}$$

or

$$\tan\theta = \frac{\sin\theta}{\cos\theta} = \frac{\left(\frac{1}{\csc\theta}\right)}{\left(\frac{1}{\sec\theta}\right)} = \frac{\left(\frac{\csc\theta\sec\theta}{\csc\theta}\right)}{\left(\frac{\csc\theta\sec\theta}{\sec\theta}\right)} = \frac{\sec\theta}{\csc\theta}$$

$$\cot\theta = \frac{\csc\theta}{\sec\theta}$$

Complementary Angle Identities

Two angles whose sum is $\pi/2$ radians (90 degrees) are *complementary*. In the diagram, the angles at vertices A and B are complementary, so we can exchange a and b, and change θ to $\pi/2 - \theta$, obtaining:

$$\sin(\pi/2-\theta) = \cos\theta$$

$$\cos(\pi/2-\theta) = \sin\theta$$

$$\tan(\pi/2-\theta) = \cot\theta$$

$$\cot(\pi/2-\theta) = \tan\theta$$

$$\sec(\pi/2-\theta) = \csc\theta$$

$$\csc(\pi/2-\theta) = \sec\theta$$

Pythagorean Identities

Identity:

$$\sin^2(x)+\cos^2(x)=1$$

The following two results follow from this and the ratio identities. To obtain the first, divide both sides of $\sin^2(x)+\cos^2(x)=1$ by $\cos^2(x)$; for the second, divide by $\sin^2(x)$.

$$\tan^2(x)+1=\sec^2(x)$$
$$1+\cot^2(x)=\csc^2(x)$$

Similarly,

$$1+\cot^2(x)=\csc^2(x)$$
$$\csc^2(x)-\cot^2(x)=1$$

Identity:

The following accounts for all three reciprocal functions:

$$\csc^2(x)+\sec^2(x)-\cot^2(x)=2+\tan^2(x)$$

Proof:

Refer to the triangle diagram. Note that $a^2+b^2=h^2$ by Pythagorean theorem.

$$\csc^2(x)+\sec^2(x)=\frac{h^2}{a^2}+\frac{h^2}{b^2}=\frac{a^2+b^2}{a^2}+\frac{a^2+b^2}{b^2}=2+\frac{b^2}{a^2}+\frac{a^2}{b^2}$$

Substituting with appropriate functions:

$$2+\frac{b^2}{a^2}+\frac{a^2}{b^2}=2+\tan^2(x)+\cot^2(x)$$

Rearranging gives:

$$\csc^2(x)+\sec^2(x)-\cot^2(x)=2+\tan^2(x)$$

Angle Sum Identities

Sine

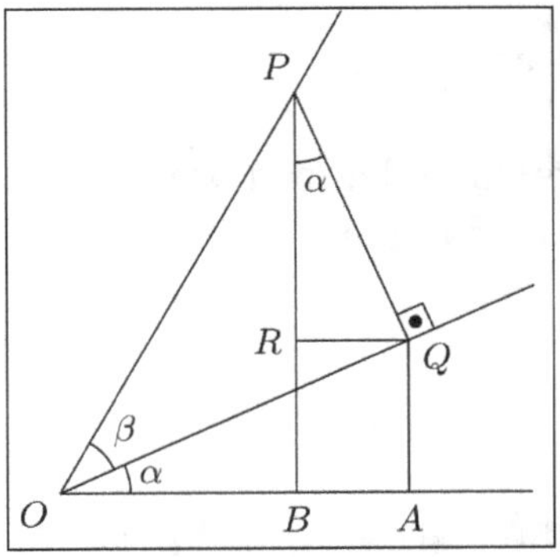

Illustration of the sum formula.

Draw a horizontal line (the x-axis); mark an origin O. Draw a line from O at an angle α above the horizontal line and a second line at an angle β above that; the angle between the second line and the x-axis is $\alpha+\beta$.

Place P on the line defined by $\alpha+\beta$ at a unit distance from the origin.

Let PQ be a line perpendicular to line OQ defined by angle α, drawn from point Q on this line to point P. $\therefore$ OQP is a right angle.

Let QA be a perpendicular from point A on the *x*-axis to Q and PB be a perpendicular from point B on the *x*-axis to P. $\therefore$ OAQ and OBP are right angles.

Draw R on PB so that QR is parallel to the x-axis.

Now angle $RPQ = \alpha$ (because $OQA = \frac{\pi}{2} - \alpha$, making

$$RQO = \alpha, RQP = \frac{\pi}{2} - \alpha, \text{ and finally } RPQ = \alpha\,)$$

$$RPQ = \frac{\pi}{2} - RQP = \frac{\pi}{2} - (\frac{\pi}{2} - RQO) = RQO = \alpha$$

$$OP = 1$$

$$PQ = \sin\beta$$

$$OQ = \cos\beta$$

$$\frac{AQ}{OQ} = \sin\alpha, \text{ so } AQ = \sin\alpha\cos\beta$$

$$\frac{PR}{PQ} = \cos\alpha, \text{ so } PR = \cos\alpha\sin\beta$$

$$\sin(\alpha+\beta) = PB = RB + PR = AQ + PR = \sin\alpha\cos\beta + \cos\alpha\sin\beta$$

By substituting $-\beta$ for β and using Symmetry, we also get:

$$\sin(\alpha-\beta) = \sin\alpha\cos(-\beta) + \cos\alpha\sin(-\beta)$$
$$\sin(\alpha-\beta) = \sin\alpha\cos\beta - \cos\alpha\sin\beta$$

Another rigorous proof, and much easier, can be given by using Euler's formula, known from complex analysis. Euler's formula is:

$$e^{i\varphi} = \cos\varphi + i\sin\varphi$$

It follows that for angles α and β we have:

$$e^{i(\alpha+\beta)} = \cos(\alpha+\beta) + i\sin(\alpha+\beta)$$

Also using the following properties of exponential functions:

$$e^{i(\alpha+\beta)} = e^{i\alpha}e^{i\beta} = (\cos\alpha + i\sin\alpha)(\cos\beta + i\sin\beta)$$

Evaluating the product:

$$e^{i(\alpha+\beta)} = (\cos\alpha\cos\beta - \sin\alpha\sin\beta) + i(\sin\alpha\cos\beta + \sin\beta\cos\alpha)$$

Equating real and imaginary parts:

$$\cos(\alpha+\beta) = \cos\alpha\cos\beta - \sin\alpha\sin\beta$$
$$\sin(\alpha+\beta) = \sin\alpha\cos\beta + \sin\beta\cos\alpha$$

Cosine

Using the figure above,

$$OP = 1$$

$$PQ = \sin\beta$$

$$OQ = \cos\beta$$

$$\frac{OA}{OQ} = \cos\alpha, \text{ so } OA = \cos\alpha\cos\beta$$

$$\frac{RQ}{PQ} = \sin\alpha, \text{ so } RQ = \sin\alpha\sin\beta$$

$$\cos(\alpha+\beta) = OB = OA - BA = OA - RQ = \cos\alpha\cos\beta - \sin\alpha\sin\beta$$

By substituting $-\beta$ for β and using Symmetry, we also get:

$$\cos(\alpha-\beta) = \cos\alpha\cos(-\beta) - \sin\alpha\sin(-\beta),$$
$$\cos(\alpha-\beta) = \cos\alpha\cos\beta + \sin\alpha\sin\beta$$

Also, using the complementary angle formulae,

$$\begin{aligned}\cos(\alpha+\beta) &= \sin(\pi/2-(\alpha+\beta)) \\ &= \sin((\pi/2-\alpha)-\beta) \\ &= \sin(\pi/2-\alpha)\cos\beta - \cos(\pi/2-\alpha)\sin\beta \\ &= \cos\alpha\cos\beta - \sin\alpha\sin\beta\end{aligned}$$

Tangent and Cotangent

From the sine and cosine formulae, we get,

$$\tan(\alpha+\beta) = \frac{\sin(\alpha+\beta)}{\cos(\alpha+\beta)} = \frac{\sin\alpha\cos\beta + \cos\alpha\sin\beta}{\cos\alpha\cos\beta - \sin\alpha\sin\beta}$$

Dividing both numerator and denominator by $\cos\alpha\cos\beta$, we get,

$$\tan(\alpha+\beta) = \frac{\tan\alpha + \tan\beta}{1 - \tan\alpha\tan\beta}$$

Subtracting β from α, using $\tan(-\beta) = -\tan\beta$,

$$\tan(\alpha-\beta) = \frac{\tan\alpha + \tan(-\beta)}{1 - \tan\alpha\tan(-\beta)} = \frac{\tan\alpha - \tan\beta}{1 + \tan\alpha\tan\beta}$$

Similarly from the sine and cosine formulae, we get,

$$\cot(\alpha+\beta)=\frac{\cos(\alpha+\beta)}{\sin(\alpha+\beta)}=\frac{\cos\alpha\cos\beta-\sin\alpha\sin\beta}{\sin\alpha\cos\beta+\cos\alpha\sin\beta}$$

Then by dividing both numerator and denominator by $\sin\alpha\sin\beta$ we get,

$$\cot(\alpha+\beta)=\frac{\cot\alpha\cot\beta-1}{\cot\alpha+\cot\beta}$$

Or, using $\cot\theta=\frac{1}{\tan\theta}$,

$$\cot(\alpha+\beta)=\frac{1-\tan\alpha\tan\beta}{\tan\alpha+\tan\beta}=\frac{\frac{1}{\tan\alpha\tan\beta}-1}{\frac{1}{\tan\alpha}+\frac{1}{\tan\beta}}=\frac{\cot\alpha\cot\beta-1}{\cot\alpha+\cot\beta}$$

Using $\cot(-\beta)=-\cot\beta$,

$$\cot(\alpha-\beta)=\frac{\cot\alpha\cot(-\beta)-1}{\cot\alpha+\cot(-\beta)}=\frac{\cot\alpha\cot\beta+1}{\cot\beta-\cot\alpha}$$

Double Angle Identities

From the angle sum identities, we get,

$$\sin(2\theta)=2\sin\theta\cos\theta \text{ and } \cos(2\theta)=\cos^2\theta-\sin^2\theta$$

The Pythagorean identities give the two alternative forms for the latter of these:

$$\cos(2\theta)=2\cos^2\theta-1$$
$$\cos(2\theta)=1-2\sin^2\theta$$

The angle sum identities also give,

$$\tan(2\theta)=\frac{2\tan\theta}{1-\tan^2\theta}=\frac{2}{\cot\theta-\tan\theta}$$
$$\cot(2\theta)=\frac{\cot^2\theta-1}{2\cot\theta}=\frac{\cot\theta-\tan\theta}{2}$$

It can also be proved using Euler's formula,

$$e^{i\varphi}=\cos\varphi+i\sin\varphi.$$

Squaring both sides yields,

$$e^{i2\varphi}=(\cos\varphi+i\sin\varphi)^2.$$

But replacing the angle with its doubled version, which achieves the same result in the left side of the equation, yields,

$$e^{i2\varphi} = \cos 2\varphi + i \sin 2\varphi.$$

It follows that,

$$(\cos\varphi + i\sin\varphi)^2 = \cos 2\varphi + i \sin 2\varphi.$$

Expanding the square and simplifying on the left hand side of the equation gives,

$$i(2\sin\varphi\cos\varphi) + \cos^2\varphi - \sin^2\varphi = \cos 2\varphi + i\sin 2\varphi,$$

because the imaginary and real parts have to be the same, we are left with the original identities,

$$\cos^2\varphi - \sin^2\varphi = \cos 2\varphi,$$

and also,

$$2\sin\varphi\cos\varphi = \sin 2\varphi.$$

Half Angle Identities

The two identities giving the alternative forms for cos 2θ lead to the following equations:

$$\cos\frac{\theta}{2} = \pm\sqrt{\frac{1+\cos\theta}{2}},$$
$$\sin\frac{\theta}{2} = \pm\sqrt{\frac{1-\cos\theta}{2}}.$$

The sign of the square root needs to be chosen properly—note that if 2π is added to θ, the quantities inside the square roots are unchanged, but the left-hand-sides of the equations change sign. Therefore, the correct sign to use depends on the value of θ.

For the tan function, the equation is:

$$\tan\frac{\theta}{2} = \pm\sqrt{\frac{1-\cos\theta}{1+\cos\theta}}.$$

Then multiplying the numerator and denominator inside the square root by (1 + cos θ) and using Pythagorean identities leads to:

$$\tan\frac{\theta}{2} = \frac{\sin\theta}{1+\cos\theta}.$$

Also, if the numerator and denominator are both multiplied by (1 - cos θ), the result is:

$$\tan\frac{\theta}{2}=\frac{1-\cos\theta}{\sin\theta}.$$

This also gives:

$$\tan\frac{\theta}{2}=\csc\theta-\cot\theta.$$

Similar manipulations for the cot function give:

$$\cot\frac{\theta}{2}=\pm\sqrt{\frac{1+\cos\theta}{1-\cos\theta}}=\frac{1+\cos\theta}{\sin\theta}=\frac{\sin\theta}{1-\cos\theta}=\csc\theta+\cot\theta.$$

Miscellaneous - The Triple Tangent Identity

If $\psi+\theta+\phi=\pi=$ half circle (for example, ψ, θ and ϕ are the angles of a triangle),

$$\tan(\psi)+\tan(\theta)+\tan(\phi)=\tan(\psi)\tan(\theta)\tan(\phi).$$

Proof:

$$\begin{aligned}\psi&=\pi-\theta-\phi\\ \tan(\psi)&=\tan(\pi-\theta-\phi)\\ &=-\tan(\theta+\phi)\\ &=\frac{-\tan\theta-\tan\phi}{1-\tan\theta\tan\phi}\\ &=\frac{\tan\theta+\tan\phi}{\tan\theta\tan\phi-1}\\ (\tan\theta\tan\phi-1)\tan\psi&=\tan\theta+\tan\phi\\ \tan\psi\tan\theta\tan\phi-\tan\psi&=\tan\theta+\tan\phi\\ \tan\psi\tan\theta\tan\phi&=\tan\psi+\tan\theta+\tan\phi\end{aligned}$$

Miscellaneous - The Triple Cotangent Identity

If $\psi+\theta+\phi=\frac{\pi}{2}=$ quarter circle,

$$\cot(\psi)+\cot(\theta)+\cot(\phi)=\cot(\psi)\cot(\theta)\cot(\phi)$$

Proof:

Replace each of ψ, θ, and ϕ with their complementary angles, so cotangents turn into tangents and vice versa.

Given:

$$\psi+\theta+\phi=\tfrac{\pi}{2}$$
$$\therefore (\tfrac{\pi}{2}-\psi)+(\tfrac{\pi}{2}-\theta)+(\tfrac{\pi}{2}-\phi)=\tfrac{3\pi}{2}-(\psi+\theta+\phi)=\tfrac{3\pi}{2}-\tfrac{\pi}{2}=\pi$$

so the result follows from the triple tangent identity.

Sum to Product Identities

$$\sin\theta\pm\sin\phi=2\sin\left(\frac{\theta\pm\phi}{2}\right)\cos\left(\frac{\theta\mp\phi}{2}\right)$$

$$\cos\theta+\cos\phi=2\cos\left(\frac{\theta+\phi}{2}\right)\cos\left(\frac{\theta-\phi}{2}\right)$$

$$\cos\theta-\cos\phi=-2\sin\left(\frac{\theta+\phi}{2}\right)\sin\left(\frac{\theta-\phi}{2}\right)$$

Proof of Sine Identities

First, start with the sum-angle identities:

$$\sin(\alpha+\beta)=\sin\alpha\cos\beta+\cos\alpha\sin\beta$$
$$\sin(\alpha-\beta)=\sin\alpha\cos\beta-\cos\alpha\sin\beta$$

By adding these together,

$$\sin(\alpha+\beta)+\sin(\alpha-\beta)=\sin\alpha\cos\beta+\cos\alpha\sin\beta+\sin\alpha\cos\beta-\cos\alpha\sin\beta=2\sin\alpha\cos\beta$$

Similarly, by subtracting the two sum-angle identities,

$$\sin(\alpha+\beta)-\sin(\alpha-\beta)=\sin\alpha\cos\beta+\cos\alpha\sin\beta-\sin\alpha\cos\beta+\cos\alpha\sin\beta=2\cos\alpha\sin\beta$$

Let $\alpha+\beta=\theta$ and $\alpha-\beta=\phi$,

$$\therefore \alpha=\frac{\theta+\phi}{2}\text{ and }\beta=\frac{\theta-\phi}{2}$$

Substitute θ and ϕ

$$\sin\theta+\sin\phi=2\sin\left(\frac{\theta+\phi}{2}\right)\cos\left(\frac{\theta-\phi}{2}\right)$$

$$\sin\theta-\sin\phi=2\cos\left(\frac{\theta+\phi}{2}\right)\sin\left(\frac{\theta-\phi}{2}\right)=2\sin\left(\frac{\theta-\phi}{2}\right)\cos\left(\frac{\theta+\phi}{2}\right)$$

Therefore,

$$\sin\theta \pm \sin\phi = 2\sin\left(\frac{\theta \pm \phi}{2}\right)\cos\left(\frac{\theta \mp \phi}{2}\right)$$

Proof of Cosine Identities

Similarly for cosine, start with the sum-angle identities:

$$\cos(\alpha+\beta) = \cos\alpha\cos\beta - \sin\alpha\sin\beta$$
$$\cos(\alpha-\beta) = \cos\alpha\cos\beta + \sin\alpha\sin\beta$$

Again, by adding and subtracting,

$$\cos(\alpha+\beta)+\cos(\alpha-\beta) = \cos\alpha\cos\beta - \sin\alpha\sin\beta + \cos\alpha\cos\beta + \sin\alpha\sin\beta = 2\cos\alpha\cos\beta$$
$$\cos(\alpha+\beta)-\cos(\alpha-\beta) = \cos\alpha\cos\beta - \sin\alpha\sin\beta - \cos\alpha\cos\beta - \sin\alpha\sin\beta = -2\sin\alpha\sin\beta$$

Substitute θ and ϕ as before,

$$\cos\theta + \cos\phi = 2\cos\left(\frac{\theta+\phi}{2}\right)\cos\left(\frac{\theta-\phi}{2}\right)$$

$$\cos\theta - \cos\phi = -2\sin\left(\frac{\theta+\phi}{2}\right)\sin\left(\frac{\theta-\phi}{2}\right)$$

Inequalities

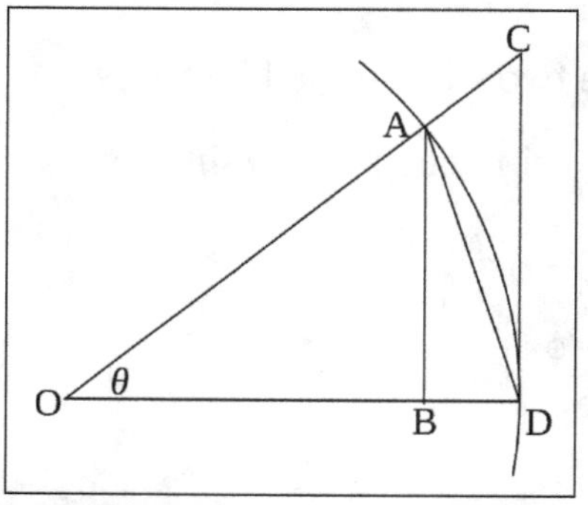

Illustration of the sine and tangent inequalities.

The figure at the right shows a sector of a circle with radius 1. The sector is $\theta/(2\pi)$ of the whole circle, so its area is $\theta/2$. We assume here that $\theta < \pi/2$.

$$OA = OD = 1$$
$$AB = \sin\theta$$
$$CD = \tan\theta$$

The area of triangle *OAD* is *AB*/2, or $\sin(\theta)/2$. The area of triangle *OCD* is *CD*/2, or $\tan(\theta)/2$.

Since triangle *OAD* lies completely inside the sector, which in turn lies completely inside triangle *OCD*, we have,

$$\sin\theta < \theta < \tan\theta.$$

This geometric argument relies on definitions of arc length and area, which act as assumptions, so it is rather a condition imposed in construction of trigonometric functions than a provable property. For the sine function, we can handle other values. If $\theta > \pi/2$, then $\theta > 1$. But $\sin\theta \leq 1$ (because of the Pythagorean identity), so $\sin\theta < \theta$. So we have,

$$\frac{\sin\theta}{\theta} < 1 \quad \text{if} \quad 0 < \theta.$$

For negative values of θ we have, by symmetry of the sine function,

$$\frac{\sin\theta}{\theta} = \frac{\sin(-\theta)}{-\theta} < 1.$$

Hence,

$$\frac{\sin\theta}{\theta} < 1 \quad \text{if} \quad \theta \neq 0,$$

and

$$\frac{\tan\theta}{\theta} > 1 \quad \text{if} \quad 0 < \theta < \frac{\pi}{2}.$$

Identities involving Calculus

Preliminaries

$$\lim_{\theta \to 0} \sin\theta = 0$$

$$\lim_{\theta \to 0} \cos\theta = 1$$

Sine and Angle Ratio Identity

$$\lim_{\theta \to 0} \frac{\sin\theta}{\theta} = 1$$

In other words, the function sine is differentiable at 0, and its derivative is 1.

Proof: From the previous inequalities, we have, for small angles,

$$\sin\theta < \theta < \tan\theta$$

Therefore,

$$\frac{\sin\theta}{\theta} < 1 < \frac{\tan\theta}{\theta}$$

Consider the right-hand inequality. Since,

$$\tan\theta = \frac{\sin\theta}{\cos\theta}$$

$$\therefore 1 < \frac{\sin\theta}{\theta\cos\theta}$$

Multiply through by $\cos\theta$

$$\cos\theta < \frac{\sin\theta}{\theta}$$

Combining with the left-hand inequality:

$$\cos\theta < \frac{\sin\theta}{\theta} < 1$$

Taking $\cos\theta$ to the limit as $\theta \to 0$

$$\lim_{\theta\to 0} \cos\theta = 1$$

Therefore,

$$\lim_{\theta\to 0} \frac{\sin\theta}{\theta} = 1$$

Cosine and Angle Ratio Identity

$$\lim_{\theta\to 0} \frac{1-\cos\theta}{\theta} = 0$$

Proof:

$$\frac{1-\cos\theta}{\theta} = \frac{1-\cos^2\theta}{\theta(1+\cos\theta)} = \frac{\sin^2\theta}{\theta(1+\cos\theta)} = \left(\frac{\sin\theta}{\theta}\right) \times \sin\theta \times \left(\frac{1}{1+\cos\theta}\right)$$

The limits of those three quantities are 1, 0, and 1/2, so the resultant limit is zero.

Cosine and Square of Angle Ratio Identity

$$\lim_{\theta\to 0} \frac{1-\cos\theta}{\theta^2} = \frac{1}{2}$$

Proof:

As in the preceding proof,

$$\frac{1-\cos\theta}{\theta^2}=\frac{\sin\theta}{\theta}\times\frac{\sin\theta}{\theta}\times\frac{1}{1+\cos\theta}.$$

The limits of those three quantities are 1, 1, and 1/2, so the resultant limit is 1/2.

Proof of Compositions of Trig and Inverse Trig Functions

All these functions follow from the Pythagorean trigonometric identity. We can prove for instance the function,

$$\sin[\arctan(x)]=\frac{x}{\sqrt{1+x^2}}$$

Proof:

We start from,

$$\sin^2\theta+\cos^2\theta=1$$

Then we divide this equation by $\cos^2\theta$.

$$\cos^2\theta=\frac{1}{\tan^2\theta+1}$$

Then use the substitution $\theta=\arctan(x)$, also use the Pythagorean trigonometric identity:

$$1-\sin^2[\arctan(x)]=\frac{1}{\tan^2[\arctan(x)]+1}$$

Then we use the identity $\tan[\arctan(x)]\equiv x$

$$\sin[\arctan(x)]=\frac{x}{\sqrt{x^2+1}}$$

HALF ANGLE FORMULA

Half Angle Formula - Sine

We start with the formula for the cosine of a double angle.

$$\cos 2\theta= 1-2\sin^2\theta$$

Now, if we let,

$$\theta = \frac{\alpha}{2}$$

then $2\theta = \alpha$ and our formula becomes:

$$\cos\alpha = 1 - 2\sin^2\left(\frac{\alpha}{2}\right)$$

We now solve for,

$$\sin\left(\frac{\alpha}{2}\right)$$

(That is, we get $\sin\left(\frac{\alpha}{2}\right)$ on the left of the equation and everything else on the right):

$$2\sin\left(\frac{}{}\right) = 1 - \cos\alpha$$

$$\sin^2\left(\frac{\alpha}{2}\right) = \frac{1-\cos}{2}$$

Solving gives us the following sine of a half-angle identity:

$$\sin\left(\frac{\alpha}{2}\right) = \pm\frac{\sqrt{1-\cos\alpha}}{2}$$

The sign (positive or negative) of $\sin\left(\frac{\alpha}{2}\right)$ depends on the quadrant in which $\frac{\alpha}{2}$ lies. If $\frac{\alpha}{2}$ is in the first or second quadrants, the formula uses the positive case:

$$\sin\left(\frac{\alpha}{2}\right) = \frac{\sqrt{1-\cos\alpha}}{2}$$

If $\frac{\alpha}{2}$ is in the third or fourth quadrants, the formula uses the negative case:

$$\sin\left(\frac{\alpha}{2}\right) = -\frac{\sqrt{1-\cos\alpha}}{2}$$

Half Angle Formula - Cosine

Using a similar process, with the same substitution of $\theta = \frac{\alpha}{2}(\text{so } 2\theta = \alpha)$ we subsitute into the identity,

$$\cos 2\theta = 2\cos^2\theta - 1$$

We obtain,

$$\cos\alpha = 2\cos^2\left(\frac{\alpha}{2}\right) - 1$$

Reverse the equation:

$$2\cos^2\left(\frac{\alpha}{2}\right) - 1 = \cos\alpha$$

Divide both sides by 2,

$$\cos^2\left(\frac{\alpha}{2}\right) = \frac{1+\cos\alpha}{2}$$

Solving for $\cos\left(\frac{\alpha}{2}\right)$, we obtain:

$$\cos\left(\frac{\alpha}{2}\right) = \pm\sqrt{\frac{1+\cos\alpha}{2}}$$

As before, the sign we need depends on the quadrant.

If $\frac{\alpha}{2}$ is in the first or fourth quadrants, the formula uses the positive case:

$$\cos\left(\frac{\alpha}{2}\right) = \sqrt{\frac{1+\cos\alpha}{2}}$$

If $\frac{\alpha}{2}$ is in the second or third quadrants, the formula uses the negative case:

$$\cos\left(\frac{\alpha}{2}\right) = -\sqrt{\frac{1+\cos\alpha}{2}}$$

Half Angle Formula - Tangent

The tangent of a half angle is given by:

$$\tan\left(\frac{\alpha}{2}\right) = \frac{1-\cos\alpha}{\sin\alpha}$$

Proof:

First, we recall $\tan x = \frac{\sin x}{\cos x}$.

$$\frac{\tan a}{2} = \frac{\frac{\sin a}{2}}{\frac{\cos a}{2}}$$

Then we use the sine and cosine of a half angle, as given above:

$$= \frac{\sqrt{\frac{1-\cos a}{2}}}{\sqrt{\frac{1+\cos a}{2}}}$$

Next line is the result of multiplying top and bottom by $\sqrt{2}$.

$$= \sqrt{\frac{1-\cos a}{1+\cos a}}$$

We then multiply top and bottom (under the square root) by $(1-\cos\alpha)$

$$= \sqrt{\frac{(1-\cos a)^2}{(1+\cos a)(1-\cos a)}}$$

Next is a difference of 2 squares.

$$= \sqrt{\frac{(1-\cos a)^2}{1-\cos^2 a}}$$

We then make use of the identity $\sin^2\theta+\cos^2\theta=1$

$$= \sqrt{\frac{(1-\cos a)^2}{\sin^2 a}}$$

We then find the square root:

$$= \frac{1-\cos a}{\sin a}$$

Of course, we would need to make allowance for positive and negative signs, depending on the quadrant in question.

We can also write the tangent of a half angle as follows:

$$\tan\left(\frac{\alpha}{2}\right) = \frac{\sin\alpha}{1+\cos\alpha}$$

Proof: We multiply numerator (top) and denominator (bottom) of the right hand side of our first result by $1+\cos\alpha$, and obtain:

$$\frac{1-\cos\alpha}{\sin\alpha}\times\frac{1+\cos\alpha}{1+\cos\alpha}$$

Next, we use the difference of 2 squares.

$$=\frac{1-\cos^2\alpha}{\sin\alpha(1+\cos\alpha)}$$

We recall $\sin^2\theta+\cos^2\theta=1$, and use it to obtain:

$$=\frac{\sin^2}{\sin\alpha(1+\cos\alpha)}$$

Finally, we cancel out the sin α.

$$=\frac{\sin\alpha}{1+\cos\alpha}$$

Tan of a Half Angle

$$\tan\left(\frac{\alpha}{2}\right)=\frac{1-\cos\alpha}{\sin\alpha}=\frac{\sin\alpha}{1+\cos\alpha}$$

Using t

It is sometimes useful to define t as the tan of a half angle:

$$t=\tan\left(\frac{\alpha}{2}\right)$$

This gives us the results:

$$\sin a=\frac{2t}{1+t^2}$$

$$\cos\alpha=\frac{1-t^2}{1+t^2}$$

$$\tan\alpha=\frac{2t}{1-t^2}$$

Tan of the Average of 2 Angles

With some algebraic manipulation, we can obtain:

$$\tan\frac{\alpha+\beta}{2}=\frac{\sin\alpha+\sin\beta}{\cos\alpha+\cos\beta}$$

Example:

Find the value 15° using the sine half-angle relationship given above.

With $\alpha=$ 30° and the formula

$$\frac{\sin\alpha}{2}=\pm\sqrt{\frac{1-\cos a}{2}}$$

we obtain:

$$\sin 15^\circ=\pm\sqrt{\frac{1-\cos 30^\circ}{2}}=\pm\sqrt{\frac{1-0.866\ldots}{2}}$$

$$=0.2588$$

Here the positive value is chosen because 15° is in the first quadrant.

Example:

Find the value of $\cos 165^\circ$ using the cosine half-angle relationship given above.

We use $\alpha=$ 330°, and $\frac{\alpha}{2}=165^\circ$.

$$\cos 165^\circ=\pm\sqrt{\frac{1+\cos 330^\circ}{2}}$$

$$=\pm\sqrt{\frac{1+0.866\ldots}{2}}$$

$$=-0.9659$$

Here the minus sign is used because 165° is in the second quadrant.

Example:

Show that $2\cos^2\left(\frac{x}{2}\right)-\cos x=1$

Using the above formula to substitute for $\frac{\cos\alpha}{2}$, we get:

$$\begin{aligned}
\text{LHS} &= 2\cos^2\left(\frac{x}{2}\right) - \cos x \\
&= 2\left(\sqrt{\frac{1+\cos x}{2}}\right)^2 - \cos x \\
&= 2\left(\frac{1+\cos x}{2}\right) - \cos x \\
&= 1 + \cos x - \cos x \\
&= 1 \\
&= \text{RHS}
\end{aligned}$$

Evaluating and Proving Half angle Identities

1. Use the half angle formula to sin 75°.

$$\begin{aligned}
\sin 75^\circ &= \pm\sqrt{\frac{1-\cos 150^\circ}{2}} \\
&= \pm\sqrt{\frac{(1+0.866)}{2}} \\
&= 0.9659
\end{aligned}$$

First Quadrant, so it's positive.

2. Find the value of $\sin\left(\frac{\alpha}{2}\right)$ if $\cos\alpha = \frac{12}{13}$ where $0^\circ < \alpha < 90^\circ$.

$$\begin{aligned}
\sin\left(\frac{\alpha}{2}\right) &= \pm\sqrt{\frac{1-\cos\alpha}{2}} \\
&= \sqrt{\frac{1-\frac{12}{13}}{2}} \\
&= \sqrt{\frac{\frac{1}{13}}{2}} \\
&= \sqrt{\frac{1}{26}} \\
&= 0.1961
\end{aligned}$$

We choose positive because we are in the first quadrant.

3. Prove the identity: $2\sin^2\left(\frac{x}{2}\right)+\cos x=1$

$$\begin{aligned}\text{LHS}&=2\sin^2\left(\frac{x}{2}\right)+\cos x\\&=2\left(\sqrt{\frac{1-\cos x}{2}}\right)^2+\cos x\\&=2\left(\frac{1-\cos x}{2}\right)+\cos x\\&=1-\cos x+\cos x\\&=1=\text{RHS}\end{aligned}$$

4. Prove the identity: $2\cos^2\left(\frac{\theta}{2}\right)\sec\theta=\sec\theta+1$

$$\begin{aligned}\text{LHS}&=2\cos^2\left(\frac{\theta}{2}\right)\sec\theta\\&=2\left(\sqrt{\frac{1+\cos\theta}{2}}\right)^2\sec\theta\\&=(1+\cos\theta)\sec\theta\\&=(1+\cos\theta)\frac{1}{\cos\theta}\\&=\sec\theta+1=\text{RHS}\end{aligned}$$

DOUBLE ANGLE FORMULAS

The double-angle formulas can be quite useful when we need to simplify complicated trigonometric expressions later.

With these formulas, it is better to remember where they come from, rather than trying to remember the actual formulas. In this way, you will understand it better and have less to clutter your memory with.

Sine of a Double Angle

$$\sin 2\alpha = 2\sin\alpha\cos\alpha$$

Proof:

The sine of the sum of two angles:

$$\sin(\alpha+\beta) = \sin\alpha\cos\beta + \cos\alpha\sin\beta$$

We will use this to obtain the sine of a double angle.

If we take the left hand side (LHS):

$$\sin(\alpha+\beta)$$

and replace β with $\boldsymbol{\alpha}$, we get:

$$\sin(\alpha+\beta) = \sin(\alpha+\alpha) = \sin 2\alpha$$

Consider the RHS:

$$\sin\alpha\cos\beta + \cos\alpha\sin\beta$$

Since we replaced β in the LHS with $\boldsymbol{\alpha}$, we need to do the same on the right side. We do this, and obtain:

$$\sin\alpha\cos\alpha + \cos\alpha\sin\alpha = 2\sin\alpha\cos\alpha$$

Putting our results for the LHS and RHS together, we obtain the important result:

$$\sin 2\alpha = 2\sin\alpha\cos\alpha$$

This result is called the sine of a double angle.

Cosine of a Double Angle

Using a similar process, we obtain the cosine of a double angle formula:

$$\cos 2\alpha = \cos^2\alpha - \sin^2\alpha$$

Proof:

This time we start with the cosine of the sum of two angles:

$$\cos(\alpha+\beta) = \cos\alpha\cos\beta - \sin\alpha\sin\beta,$$

and once again replace β with α on both the LHS and RHS, as follows:

$$\text{LHS} = \cos(\alpha+\alpha) = \cos(2\alpha)$$

$$\text{RHS} = \cos\alpha\cos\alpha - \sin\alpha\sin\alpha = \cos^2\alpha - \sin^2\alpha$$

Different Forms of the Cosine Double Angle Result

By using the result $\sin^2 \alpha + \cos^2 \alpha = 1$, we can write the RHS of the above formula as:

$$\cos^2 \alpha - \sin^2 \alpha$$
$$= \left(1 - \sin^2 \alpha\right) - \sin^2 \alpha$$
$$= 1 - 2\sin^2 \alpha$$

Likewise, we can substitute $(1 - \cos^2 \alpha)$ for $\sin^2 \alpha$ into our RHS and obtain:

$$\cos^2 \alpha - \sin^2 \alpha$$
$$= \cos^2 \alpha - (1 - \cos^2 \alpha)$$
$$= 2\cos^2 \alpha - 1$$

The following have equivalent value, and we can use whichever one we like, depending on the situation:

$$\cos 2\alpha = \cos^2 \alpha - \sin^2 \alpha$$
$$\cos 2\alpha = 1 - 2\sin^2 \alpha$$
$$\cos 2\alpha = 2\cos^2 \alpha - 1$$

Example: Find cos 60° by using the functions of 30°.

Answer: We can only use the sine and cosine functions of 30°, so we need to start with 60°=2 x 30°.

We will use the result:

$$\cos 2\alpha = \cos^2 \alpha - \sin^2 \alpha$$

and the 30-60 triangle:

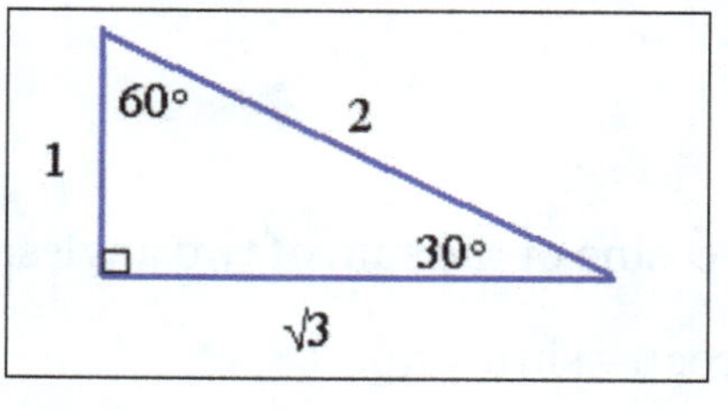

Now we proceed to find the exact value of cos 60° using the ratios of 30°:

$$\cos 60^\circ = \cos\left(2 \times 30^\circ\right)$$
$$= \cos^2\left(30^\circ\right) - \sin^2\left(30^\circ\right)$$

$$=\left(\frac{\sqrt{3}}{2}\right)^2-\left(\frac{1}{2}\right)^2$$

$$=\frac{3}{4}-\frac{1}{4}$$

$$=\frac{2}{4}$$

$$=\frac{1}{2}$$

Example: Find the exact value of $\cos 2x$ if $\sin x=-\frac{12}{13}$ (in Quadrant III).

Answer: Using the following form of the cosine of a double angle formula,

$$\cos 2\alpha = 1- 2\sin^2\alpha ,$$

we have:

$$\cos 2x=1-2\sin^2 x$$

$$=1-2\left(\frac{-12}{13}\right)^2$$

$$=1-2\left(\frac{144}{169}\right)$$

$$=\frac{169-288}{169}$$

$$=\frac{-119}{169}$$

Example: Without finding x, find the exact value of $\sin 2x$ if $\cos x=\frac{4}{5}$ (in Quadrant I).

Answer: We recognise that we need to use the 3-4-5 triangle.

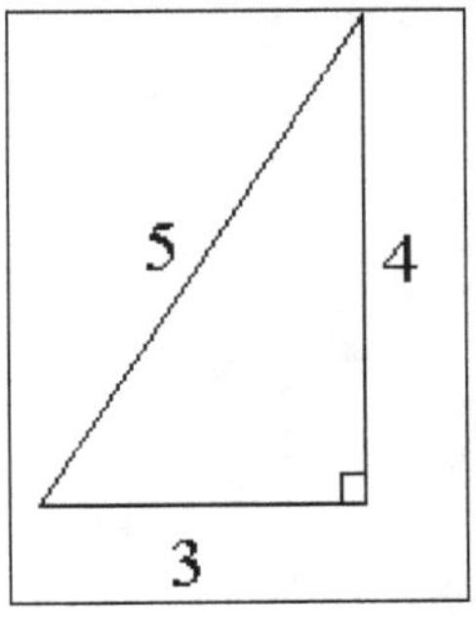

We can use our formula for the sine of a double angle to find the required value:

$$\begin{aligned}\sin 2x &= 2\sin x\cos x\\ &= 2\left(\frac{3}{5}\right)\left(\frac{4}{5}\right)\\ &= \frac{24}{25}\end{aligned}$$

Prove that: $\dfrac{1-\tan^2 x}{\sec^2 x} = \cos 2x$

Answer: When we need to prove an identity, we start on one side and work on it until it is equivalent to the other side.

In this example, we start on the left hand side and use our various identities from earlier sections to simplify it.

$$\begin{aligned}\text{LHS} &= \frac{1-\tan^2 x}{\sec^2 x}\\ &= \frac{1}{\sec^2 x}\left(1-\tan^2 x\right)\\ &= \cos^2 x\left(1-\frac{\sin^2 x}{\cos^2 x}\right)\\ &= \cos^2 x \times \frac{\cos^2 x-\sin^2 x}{\cos^2 x}\\ &= \cos^2 x-\sin^2 x\\ &= \cos 2x\\ &= \text{RHS}\end{aligned}$$

Prove that: $2\csc 2x\tan x = \sec^2$

Answer: $$\begin{aligned}\text{LHS} &= 2\csc 2x\tan x\\ &= \frac{2\tan x}{\sin 2x}\\ &= \frac{2\tan x}{2\sin x\cos x}\\ &= \frac{\sin x}{\cos x}\times\frac{1}{\sin x\cos x}\\ &= \frac{1}{\cos^2 x}\\ &= \sec^2 x\\ &= \text{RHS}\end{aligned}$$

TRIPLE ANGLE FORMULAS

Triple Angle Formulas/Sine

Theorem:

$$\sin 3\theta = 3\sin\theta - 4\sin^3\theta$$

where sin denotes sine.

Proof:

$$\begin{aligned}
\sin 3\theta &= \sin(2\theta+\theta) \\
&= \sin 2\theta\cos\theta + \cos 2\theta\sin\theta && \text{Sine of Sum} \\
&= (2\sin\theta\cos\theta)\cos\theta + (\cos^2\theta - \sin^2\theta)\sin\theta && \text{Double Angle Formula for Sine and Double Angle Formula for Cosine} \\
&= 2\sin\theta\cos^2\theta + \cos^2\theta\sin\theta - \sin^3\theta \\
&= 2\sin\theta(1-\sin^2\theta) + (1-\sin^2\theta)\sin\theta - \sin^3\theta && \text{Sum of Squares of Sine and Cosine} \\
&= 2\sin\theta - 2\sin^3\theta + \sin\theta - \sin^3\theta - \sin^3\theta && \text{Multiplying out} \\
&= 3\sin\theta - 4\sin^3\theta && \text{gathering terms}
\end{aligned}$$

Proof:

$$\begin{aligned}
\cos 3\theta + i\sin 3\theta &= (\cos\theta + i\sin\theta)^3 && \text{De Moivre's Formula} \\
&= (\cos\theta)^3 + \binom{3}{1}(\cos\theta)^2(i\sin\theta) + \binom{3}{2}(\cos\theta)(i\sin\theta)^2 + (i\sin\theta)^3 && \text{Binomial Theorem} \\
&= \cos^3\theta + 3i\cos^2\theta\sin\theta + 3i^2\cos\theta\sin^2\theta + i^3\sin^3\theta && \text{Substituting for binomial coefficients} \\
&= \cos^3\theta + 3i\cos^2\theta\sin\theta - 3\cos\theta\sin^2\theta - i\sin^3\theta && i^2 = -1 \\
&= \cos^3\theta - 3\cos\theta\sin^2\theta + i(3\cos^2\theta\sin\theta - \sin^3\theta) && \text{Rearranging}
\end{aligned}$$

Hence:

$$\begin{aligned}
\sin 3\theta &= 3\cos^2\theta\sin\theta - \sin^3\theta && \text{Equating in previous equation} \\
&= 3(1-\sin^2\theta)\sin\theta - \sin^3\theta && \text{Sum of Squares of Sine and Cosine} \\
&= 3\sin\theta - 4\sin^3\theta && \text{Multiplying out and gathering terms}
\end{aligned}$$

Triple Angle Formulas/Cosine

Theorem:

$$\cos 3\theta = 4\cos^3\theta - 3\cos\theta$$

where cos denotes cosine.

Example: $2\cos 3\theta + 1$

$$2\cos 3\theta + 1 = \left(\cos\theta - \cos\frac{2\pi}{9}\right)\left(\cos\theta - \cos\frac{4\pi}{9}\right)\left(\cos\theta - \cos\frac{8\pi}{9}\right)$$

Proof:

$\cos 3\theta = \cos(2\theta + \theta)$	
$= \cos 2\theta\cos\theta - \sin 2\theta\sin\theta$	Cosine of Sum
$= (\cos^2\theta - \sin^2\theta)\cos\theta - (2\sin\theta\cos\theta)\sin\theta$	Double Angle Formula for Cosine and Double Angle Formula for Sine
$= \cos^3\theta - \sin^2\theta\cos\theta - 2\sin^2\theta\cos\theta$	
$= \cos^3\theta - (1 - \cos^2\theta)\cos\theta - 2(1 - \cos^2\theta)\cos\theta$	Sum of Squares of Sine and Cosine
$= \cos^3\theta - \cos\theta + \cos^3\theta - 2\cos\theta + 2\cos^3\theta$	Multiplying out
$= 4\cos^3\theta - 3\cos\theta$	gathering terms

Proof:

We have:

$\cos 3\theta + i\sin 3\theta = (\cos\theta + i\sin\theta)^3$	De Moivre's Formula
$= (\cos\theta)^3 + \binom{3}{1}(\cos\theta)^2(i\sin\theta) + \binom{3}{2}(\cos\theta)(i\sin\theta)^2 + (i\sin\theta)^3$	Binomial Theorem
$= \cos^3\theta + 3i\cos^2\theta\sin\theta + 3i^2\cos\theta\sin^2\theta + i^3\sin^3\theta$	Substituting for binomial coefficients
$= \cos^3\theta + 3i\cos^2\theta\sin\theta - 3\cos\theta\sin^2\theta - i\sin^3\theta$	$i^2 = -1$
$= \cos^3\theta - 3\cos\theta\sin^2\theta + i(3\cos^2\theta\sin\theta - \sin^3\theta)$	Rearranging

Hence:

$\cos 3\theta = \cos^3\theta - 3\cos\theta\sin^2\theta$	Equating real parts in previous equation
$= \cos^3\theta - 3\cos\theta(1 - \cos^2\theta)$	Sum of Squares of Sine and Cosine
$= 4\cos^3\theta - 3\cos\theta$	Multiplying out and gathering terms

Proof:

$\cos 3\theta = \cos(2\theta + \theta)$	
$= \cos 2\theta\cos\theta - \sin 2\theta\sin\theta$	Cosine of Sum
$= (2\cos^2\theta - 1)\cos\theta - 2\sin^2\theta\cos\theta$	Double Angle Formula for Cosine and Double Angle Formula for Sine
$= (2\cos^2\theta - 1 - 2(1 - 2\cos^2\theta))\cos\theta$	Sum of Squares of Sine and Cosine
$= 4\cos^3\theta - 3\cos\theta$	Gathering terms

Triple Angle Formulas/Tangent

Theorem:

$$\tan(3\theta) = \frac{3\tan\theta - \tan^3\theta}{1 - 3\tan^2\theta}$$

where tan denotes tangent.

Proof:

$$\tan(3\theta) = \frac{\sin(3\theta)}{\cos(3\theta)}$$ Tangent is Sine divided by Cosine

$$= \frac{3\sin\theta - 4\sin^3\theta}{4\cos^3\theta - 3\cos\theta}$$ Triple Angle Formula for Sine and Triple Angle Formula for Cosine

$$= \frac{3\sin\theta - 4\sin^3\theta}{4\cos^3\theta - 3\cos\theta} \frac{\cos^3\theta}{\cos^3\theta}$$

$$= \frac{\frac{3\tan\theta}{\cos^2\theta} - 4\tan^3\theta}{4 - \frac{3\cos\theta}{\cos^3\theta}}$$ Tangent is Sine divided by Cosine

$$= \frac{3\tan\theta\sec^2\theta - 4\tan^3\theta}{4 - 3\sec^2\theta}$$ Secant is Reciprocal of Cosine

$$= \frac{3\tan\theta\left(1 + \tan^2\theta\right) - 4\tan^3\theta}{4 - 3\left(1 + \tan^2\theta\right)}$$ Difference of Squares of Secant and Tangent

$$= \frac{3\tan\theta + 3\tan^3\theta - 4\tan^3\theta}{4 - 3 - 3\tan^2\theta}$$ Multiplying out

$$= \frac{3\tan\theta - \tan^3\theta}{1 - 3\tan^2\theta}$$ Gathering terms

Proof:

$$\tan(3\theta) = \frac{\tan\theta + \tan(2\theta)}{1 - \tan\theta\tan(2\theta)}$$ Tangent of Sum

$$= \frac{\tan\theta + \frac{2\tan\theta}{1 - \tan^2\theta}}{1 - \tan\theta\frac{2\tan\theta}{1 - \tan^2\theta}}$$ Double Angle Formula for Tangent

$$= \frac{\tan\theta\left(1 - \tan^2\theta\right) + 2\tan\theta}{\left(1 - \tan^2\theta\right) - 2\tan^2\theta}$$ Simplifying

$$= \frac{\tan\theta - \tan^3\theta + 2\tan\theta}{1 - \tan^2\theta - 2\tan^2\theta} \quad \text{Simplifying}$$

$$= \frac{3\tan\theta - \tan^3\theta}{1 - 3\tan^2\theta} \quad \text{Simplifying}$$

References

- Trigidentities: mathopenref.com, Retrieved 01 May, 2020
- 3-double-angle-formulas, analytic-trigonometry: intmath.com, Retrieved 14 February, 2020
- Tangent, Triple-Angle-Formulas: proofwiki.org, Retrieved 16 June, 2020
- Thomas W. Hungerford, Douglas J. Shaw (2008). "§6.2 The sine, cosine and tangent functions". Contemporary Precalculus: A Graphing Approach (5th ed.). Cengage Learning. p. 442. ISBN 0-495-10833-2
- 4-half-angle-formulas, analytic-trigonometry: intmath.com, Retrieved 05 March, 2020

5

Trigonometry Laws

There are various laws that are studied under the domain of trigonometry. These are law of sines, cosines which includes spherical and hyperbolic law of cosines, and tangents. This chapter has been carefully written to provide an easy understanding of these trigonometry laws.

LAW OF SINES

The Law of Sines is the relationship between the sides and angles of non-right (oblique) triangles. Simply, it states that the ratio of the length of a side of a triangle to the sine of the angle opposite that side is the same for all sides and angles in a given triangle.

In ΔABC is an oblique triangle with sides a,b and c, then:

$$\frac{a}{\sin A} = \frac{b}{\sin B} = \frac{c}{\sin C}.$$

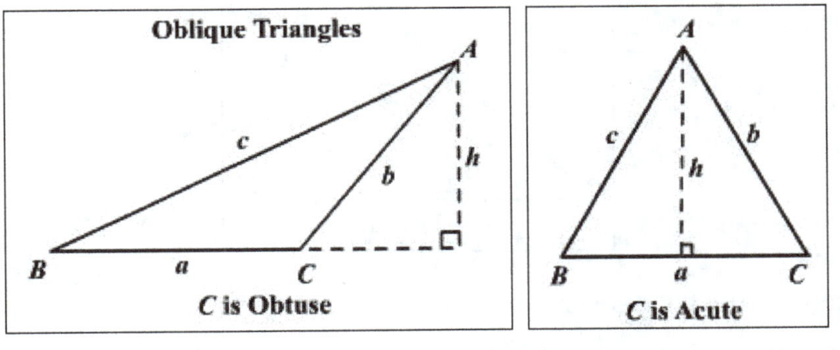

To use the Law of Sines you need to know either two angles and one side of the triangle (AAS or ASA) or two sides and an angle opposite one of them (SSA). Notice that for the first two cases we use the same parts that we used to prove congruence of triangles in geometry but in the last case we could not prove congruent triangles given these parts. This is because the remaining pieces could have been different sizes.

Example: Given two angles and a non-included side (AAS).

Given ΔABC with $m\angle A = 30°$, $m\angle B = 20°$ and a = 45 m. Find the remaining angle and sides.

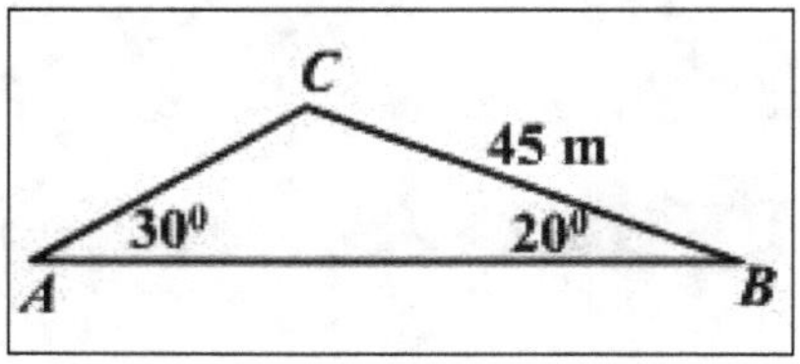

The third angle of the triangle is,

$$m\angle C = 180° - m\angle A - m\angle B = 180° - 30° - 20° = 130°$$

By the Law of Sines,

$$\frac{45}{\sin 30°} = \frac{b}{\sin 20°} = \frac{c}{\sin 130°}$$

By the Properties of Proportions,

$$b = \frac{45 \sin 20°}{\sin 30°} \approx 30.78\text{m and } c = \frac{45 \sin 130°}{\sin 30°} \approx 68.94\text{m}$$

Example: Given two angles and an included side (ASA).

Given $m\angle A = 42°$, $m\angle B = 75°$, and c = 22 cm. Find the remaining angle and sides.

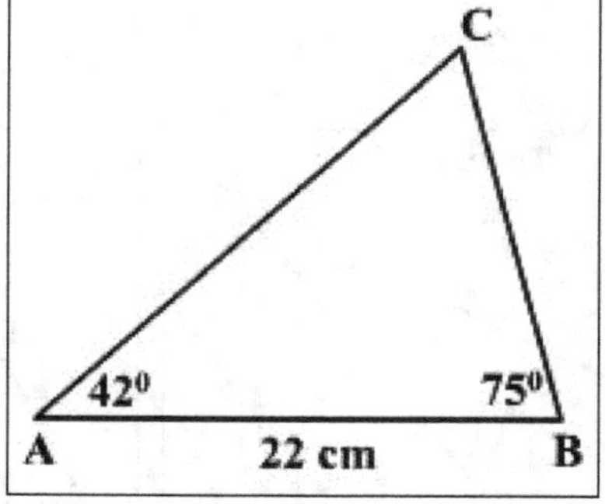

The third angle of the triangle is:

$$m\angle C = 180° - m\angle A - m\angle B = 180° - 42° - 75° = 63°$$

By the Law of Sines,

$$\frac{a}{\sin 42°} = \frac{b}{\sin 75°} = \frac{22}{\sin 63°}$$

By the Properties of Proportions

$$a = \frac{22\sin 42^\circ}{\sin 63^\circ} \approx 16.52 \text{ cm} \text{ and } b = \frac{22\sin 75^\circ}{\sin 63^\circ} \approx 23.85 \text{ cm}$$

The Ambiguous Case

If two sides and an angle opposite one of them are given, three possibilities can occur.

1. No such triangle exists.
2. Two different triangles exist.
3. Exactly one triangle exists.

Consider a triangle in which you are given a, b and A. (The altitude h from vertex B to side $\overline{AC}$, by the definition of sines is equal to b sin A).

1. No such triangle exists if A is acute and a < h or A is obtuse and a ≤ b.

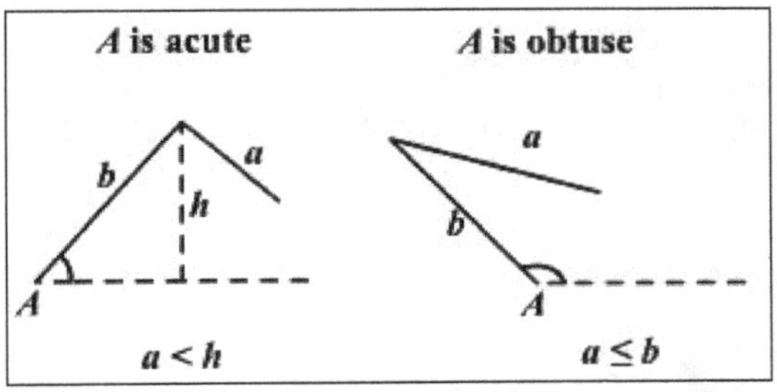

2. Two different triangles exist if A is acute and h < a < b.

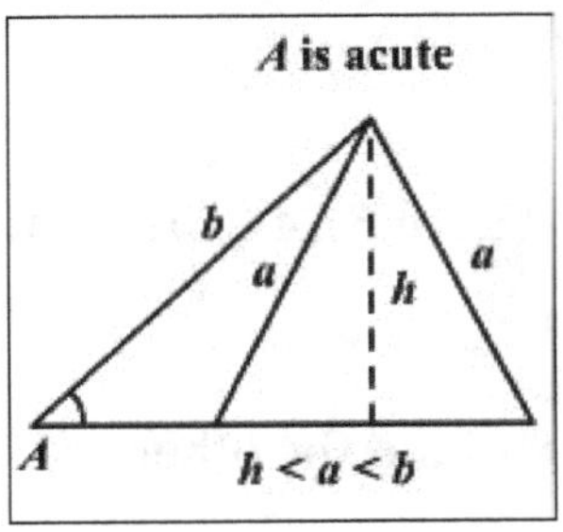

3. In every other case, exactly one triangle exists.

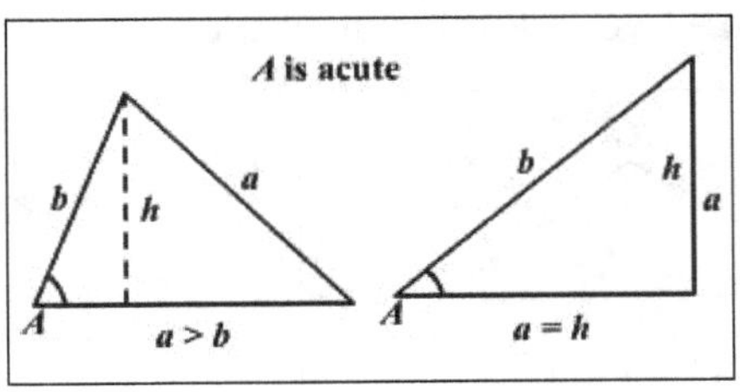

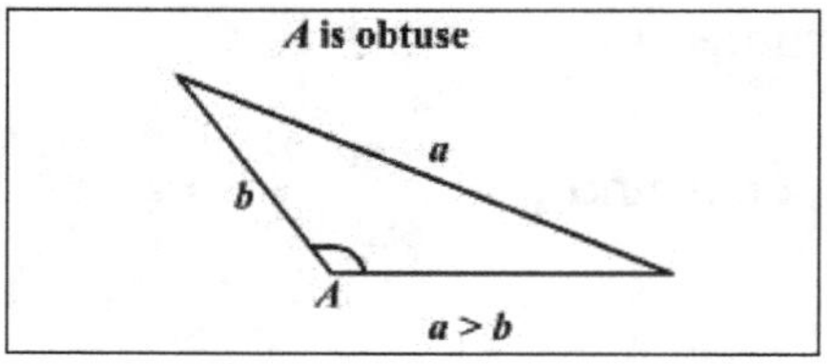

Example: No Solution Exists

Given $a = 15$, $b = 25$ and $m\angle A = 80°$. Find the other angles and side.

$$h = b \sin A = 25 \sin 80° \approx 24.6$$

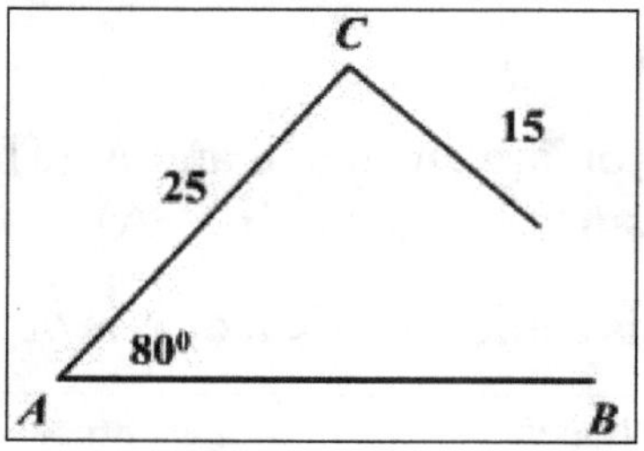

Notice that a < h. So it appears that there is no solution. Verify this using the Law of Sines.

$$\frac{a}{\sin A} = \frac{b}{\sin B}$$

$$\frac{15}{\sin 80°} = \frac{25}{\sin B}$$

$$\sin B = \frac{25 \sin 80°}{15} \approx 1.641 > 1$$

This contrasts the fact that the $-1 \le \sin B \le 1$. Therefore, no triangle exists.

Example: Two Solutions Exist.

Given $a = 6, b = 7$ and $m\angle A = 30°$. Find the other angles and side.

$$h = b \sin A = 7 \sin 30° = 3.5$$

$h < a < b$ therefore, there are two triangles possible.

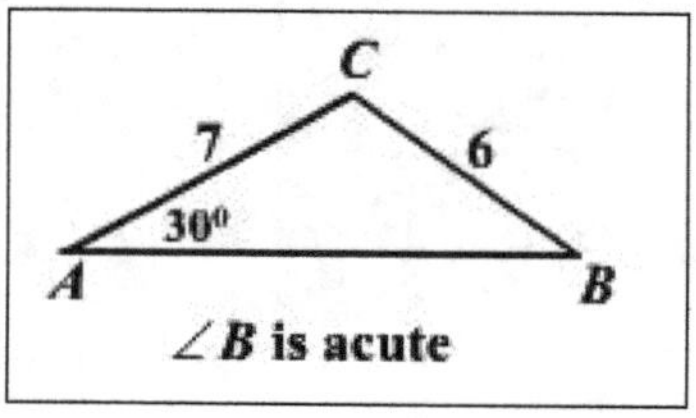

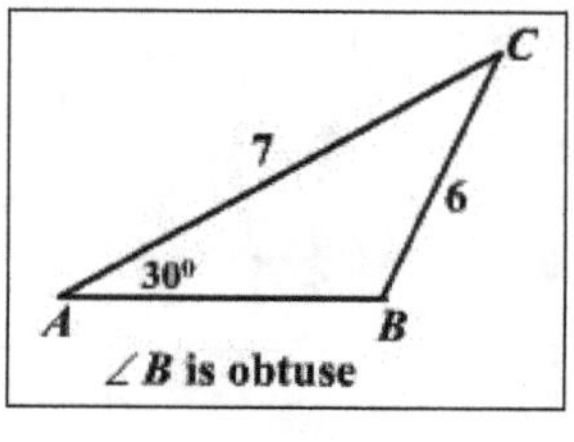

By the Law of Sines,

$$\frac{a}{\sin A} = \frac{b}{\sin B}$$

$$\sin B = \frac{b \sin A}{a} = \frac{7 \sin 30°}{6} \approx 0.5833$$

There are two angles between 0° and 180° whose sine is approximately 0.5833, are 35.69° and 144.31°.

If $B \approx 35.69°$

$C \approx 180° - 30° - 35.69° = 114.31°$

$$c = \frac{a \sin C}{\sin A} \approx \frac{6 \sin 114.31°}{\sin 30°} \approx 10.94$$

If $B \approx 144.31°$

$C \approx 180° - 30° - 144.31° = 5.69°$

$$c \approx \frac{6 \sin 5.69°}{\sin 30°} \approx 1.19$$

Example: One Solution Exists.

Given $a = 22, b = 12$ and $m\angle A = 40°$. Find the other angles and side.

$a > b$

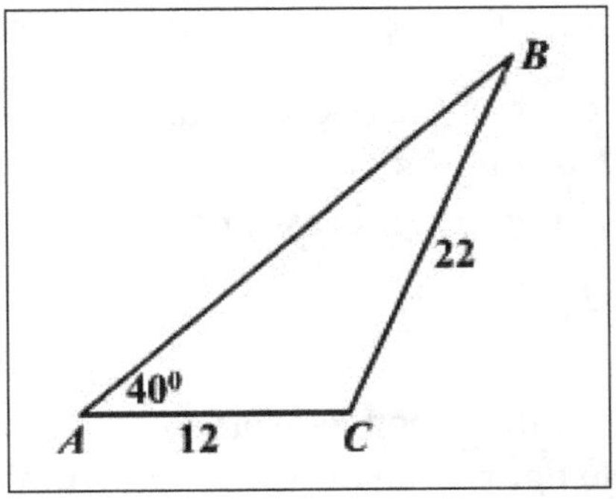

By the Law of Sines,

$$\frac{a}{\sin A} = \frac{b}{\sin B}$$

$$\sin B = \frac{b \sin A}{a} = \frac{12 \sin 40°}{22} \approx 0.3506$$

$B \approx 20.52°$

B is acute.

$$m\angle C = 180° - m\angle A - m\angle B = 180° - 40° - 20.52° = 29.79°$$

By the Law of Sines,

$$\frac{c}{\sin 119.48^\circ} = \frac{22}{\sin 40^\circ}$$

$$c = \frac{22 \sin 119.48^\circ}{\sin 40^\circ} \approx 29.79$$

LAW OF COSINES

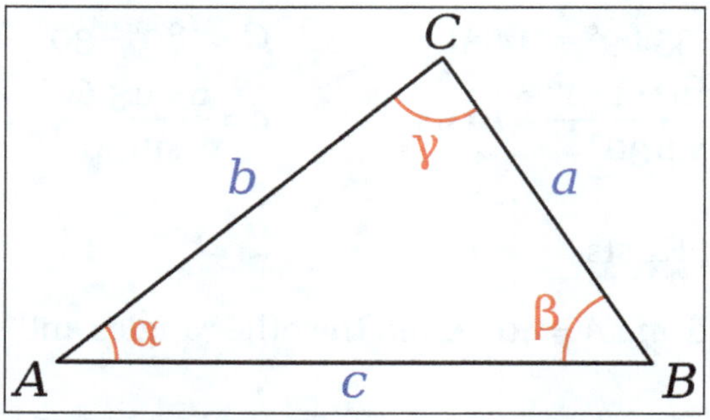

A triangle: The angles α (or A), β (or B), and γ (or C) are respectively opposite the sides a, b, and c.

In trigonometry, the law of cosines (also known as the cosine formula, cosine rule, or al-Kashi's theorem) relates the lengths of the sides of a triangle to the cosine of one of its angles. Using notation as in figure, the law of cosines states:

$$c^2 = a^2 + b^2 - 2ab\cos\gamma,$$

where γ denotes the angle contained between sides of lengths a and b and opposite the side of length c. For the same figure, the other two relations are analogous:

$$a^2 = b^2 + c^2 - 2bc\cos\alpha,$$

$$b^2 = a^2 + c^2 - 2ac\cos\beta.$$

The law of cosines generalizes the Pythagorean theorem, which holds only for right triangles: if the angle γ is a right angle (of measure 90 degrees, or $\frac{\pi}{2}$ radians), then $\cos\gamma = 0$, and thus the law of cosines reduces to the Pythagorean theorem:

$$c^2 = a^2 + b^2.$$

The law of cosines is useful for computing the third side of a triangle when two sides and their enclosed angle are known, and in computing the angles of a triangle if all three sides are known.

Applications

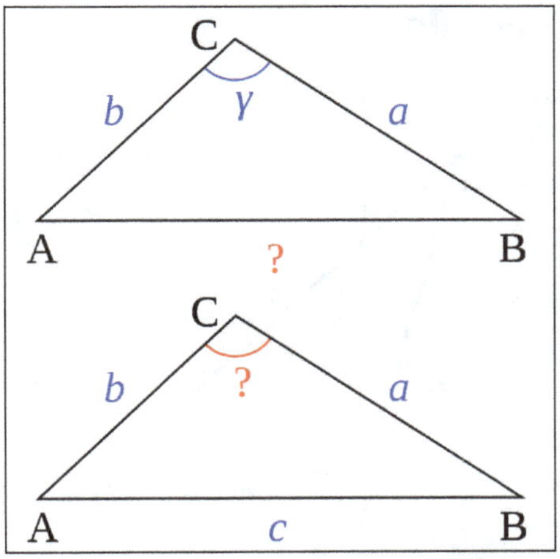

Applications of the law of cosines: unknown side and unknown angle.

The theorem is used in triangulation, for solving a triangle or circle, i.e., to find:

- The third side of a triangle if one knows two sides and the angle between them:

$$c = \sqrt{a^2 + b^2 - 2ab\cos\gamma};$$

- The angles of a triangle if one knows the three sides:

$$\gamma = \arccos\left(\frac{a^2 + b^2 - c^2}{2ab}\right);$$

- The third side of a triangle if one knows two sides and an angle opposite to one of them (one may also use the Pythagorean theorem to do this if it is a right triangle):

$$a = b\cos\gamma \pm \sqrt{c^2 - b^2\sin^2\gamma}.$$

These formulas produce high round-off errors in floating point calculations if the triangle is very acute, i.e., if c is small relative to a and b or γ is small compared to 1. It is even possible to obtain a result slightly greater than one for the cosine of an angle.

The third formula shown is the result of solving for a in the quadratic equation $a^2 - 2ab \cos \gamma + b^2 - c^2 = 0$. This equation can have 2, 1, or 0 positive solutions corresponding to the number of possible triangles given the data. It will have two positive solutions if $b \sin \gamma < c < b$, only one positive solution if $c = b \sin \gamma$, and no solution if $c < b \sin \gamma$ or $c \geq b$. These different cases are also explained by the side-side-angle congruence ambiguity.

Proofs:

Using the Distance Formula

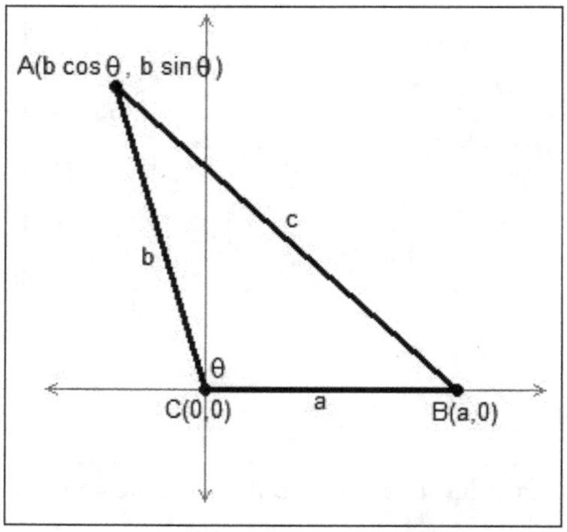

Coordinate geometry proof.

Consider a triangle with sides of length *a*, *b*, *c*, where θ is the measurement of the angle opposite the side of length *c*. This triangle can be placed on the Cartesian coordinate system by plotting the following points:

$$A = (b\cos\theta, b\sin\theta), B = (a, 0), \text{ and } C = (0, 0).$$

By the distance formula, we have,

$$c = \sqrt{(a - b\cos\theta)^2 + (0 - b\sin\theta)^2}.$$

Now, we just work with that equation:

$$c^2 = (a - b\cos\theta)^2 + (-b\sin\theta)^2$$
$$c^2 = a^2 - 2ab\cos\theta + b^2\cos^2\theta + b^2\sin^2\theta$$
$$c^2 = a^2 + b^2(\sin^2\theta + \cos^2\theta) - 2ab\cos\theta$$
$$c^2 = a^2 + b^2 - 2ab\cos\theta.$$

An advantage of this proof is that it does not require the consideration of different cases for when the triangle is acute, right, or obtuse.

Using Trigonometry

Dropping the perpendicular onto the side *c* through point *C*, an altitude of the triangle, shows:

$$c = a\cos\beta + b\cos\alpha.$$

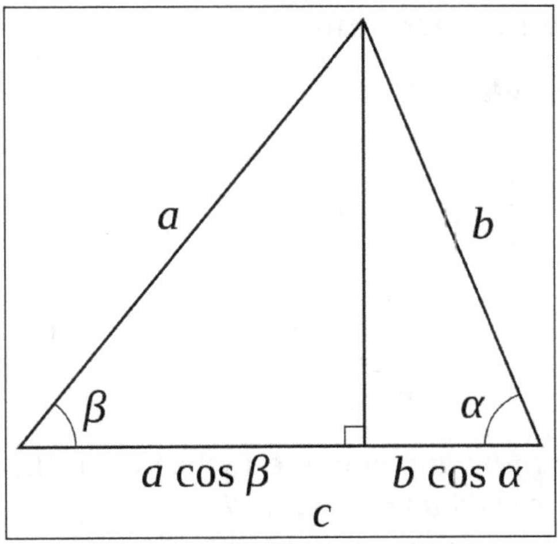

An acute triangle with perpendicular.

(This is still true if α or β is obtuse, in which case the perpendicular falls outside the triangle.) Multiplying through by *c* yields,

$$c^2 = ac\cos\beta + bc\cos\alpha.$$

Considering the two other altitudes of the triangle yields,

$$a^2 = ac\cos\beta + ab\cos\gamma,$$
$$b^2 = bc\cos\alpha + ab\cos\gamma.$$

Adding the latter two equations gives,

$$a^2 + b^2 = ac\cos\beta + bc\cos\alpha + 2ab\cos\gamma.$$

Subtracting the first equation from the last one results in,

$$a^2 + b^2 - c^2 = ac\cos\beta + bc\cos\alpha + 2ab\cos\gamma - (ac\cos\beta + bc\cos\alpha)$$

which simplifies to,

$$c^2 = a^2 + b^2 - 2ab\cos\gamma.$$

This proof uses trigonometry in that it treats the cosines of the various angles as quantities in their own right. It uses the fact that the cosine of an angle expresses the relation between the two sides enclosing that angle in *any* right triangle. Other proofs are more geometric in that they treat an expression such as $a \cos \gamma$ merely as a label for the length of a certain line segment.

Many proofs deal with the cases of obtuse and acute angles γ separately.

Using the Pythagorean Theorem

Case of an Obtuse Angle

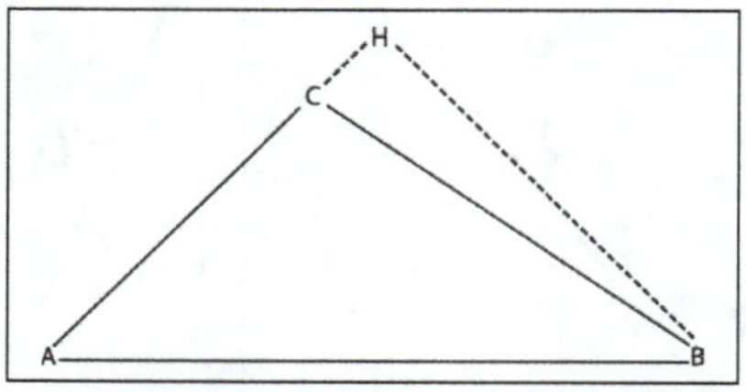

Obtuse triangle *ABC* with height *BH*.

Euclid proves this theorem by applying the Pythagorean theorem to each of the two right triangles in the figure shown (*AHB* and *CHB*). Using *d* to denote the line segment *CH* and *h* for the height *BH*, triangle *AHB* gives us,

$$c^2 = (b+d)^2 + h^2,$$

and triangle *CHB* gives,

$$d^2 + h^2 = a^2.$$

Expanding the first equation gives,

$$c^2 = b^2 + 2bd + d^2 + h^2.$$

Substituting the second equation into this, the following can be obtained:

$$c^2 = a^2 + b^2 + 2bd.$$

To transform it into the modern form of the law of cosines, note that,

$$d = a\cos(\pi - \gamma) = -a\cos\gamma.$$

Case of an Acute Angle

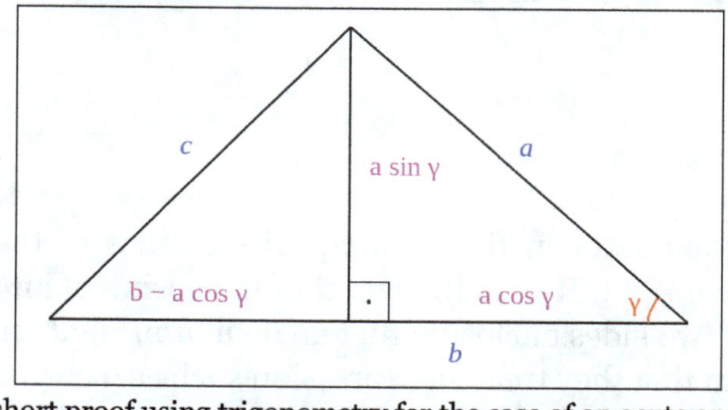

A short proof using trigonometry for the case of an acute angle.

Euclid's proof of his Proposition 13 proceeds along the same lines as his proof of Proposition 12: he applies the Pythagorean theorem to both right triangles formed by dropping the perpendicular onto one of the sides enclosing the angle γ and uses the binomial theorem to simplify.

Another Proof in the Acute Case

Using more trigonometry, the law of cosines can be deduced by using the Pythagorean theorem only *once*. In fact, by using the right triangle on the left hand side of figure it can be shown that:

$$\begin{aligned} c^2 &= (b - a\cos\gamma)^2 + (a\sin\gamma)^2 \\ &= b^2 - 2ab\cos\gamma + a^2\cos^2\gamma + a^2\sin^2\gamma \\ &= b^2 + a^2 - 2ab\cos\gamma, \end{aligned}$$

using the trigonometric identity,

$$\cos^2\gamma + \sin^2\gamma = 1.$$

This proof needs a slight modification if $b < a\cos(\gamma)$. In this case, the right triangle to which the Pythagorean theorem is applied moves *outside* the triangle *ABC*. The only effect this has on the calculation is that the quantity $b - a\cos(\gamma)$ is replaced by $a\cos(\gamma) - b$. As this quantity enters the calculation only through its square, the rest of the proof is unaffected. However, this problem only occurs when β is obtuse, and may be avoided by reflecting the triangle about the bisector of γ.

Referring to figure it is worth noting that if the angle opposite side a is α then:

$$\tan\alpha = \frac{a\sin\gamma}{b - a\cos\gamma}.$$

This is useful for direct calculation of a second angle when two sides and an included angle are given.

Using Ptolemy's Theorem

Referring to the diagram, triangle *ABC* with sides $AB = c$, $BC = a$ and $AC = b$ is drawn inside its circumcircle as shown. Triangle *ABD* is constructed congruent to triangle *ABC* with $AD = BC$ and $BD = AC$. Perpendiculars from *D* and *C* meet base *AB* at *E* and *F* respectively. Then:

$$\begin{aligned} &BF = AE = BC\cos\hat{B} = a\cos\hat{B} \\ &\Rightarrow DC = EF = AB - 2BF = c - 2a\cos\hat{B}. \end{aligned}$$

Now the law of cosines is rendered by a straightforward application of Ptolemy's theorem to cyclic quadrilateral *ABCD*:

$$AD \times BC + AB \times DC = AC \times BD$$

$$\Rightarrow a^2 + c(c - 2a\cos\hat{B}) = b^2$$

$$\Rightarrow a^2 + c^2 - 2ac\cos\hat{B} = b^2.$$

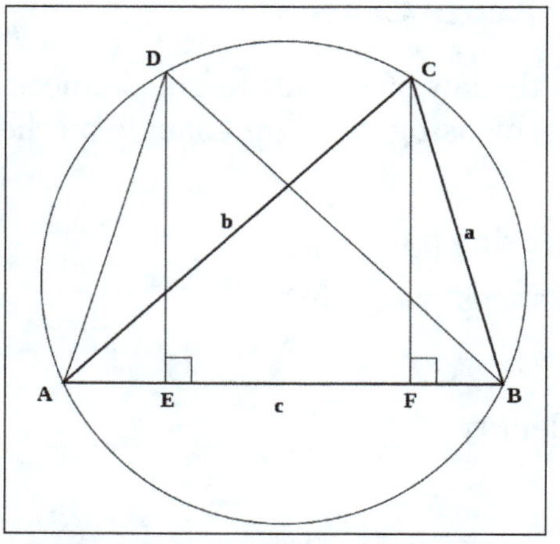

Proof of law of cosines using Ptolemy's theorem.

Plainly if angle *B* is right, then *ABCD* is a rectangle and application of Ptolemy's theorem yields the Pythagorean theorem:

$$a^2 + c^2 = b^2.$$

By Comparing Areas

One can also prove the law of cosines by calculating areas. The change of sign as the angle γ becomes obtuse makes a case distinction necessary.

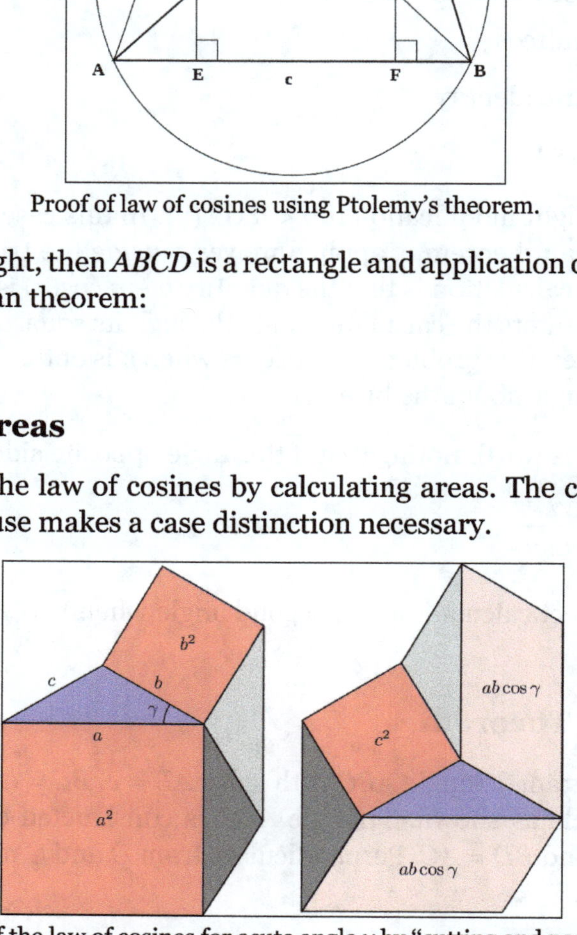

Proof of the law of cosines for acute angle γ by "cutting and pasting".

Recall that:

- a^2, b^2, and c^2 are the areas of the squares with sides *a*, *b*, and *c*, respectively.
- If γ is acute, then ab cos γ is the area of the parallelogram with sides a and b forming an angle of $\gamma' = \frac{\pi}{2} - \gamma$.

- If γ is obtuse, and so $\cos \gamma$ is negative, then $-ab \cos \gamma$ is the area of the parallelogram with sides a and b forming an angle of $\gamma' = \gamma - \frac{\pi}{2}$.

Acute Case

Figure shows a heptagon cut into smaller pieces (in two different ways) to yield a proof of the law of cosines. The various pieces are:

- In pink, the areas a^2, b^2 on the left and the areas $2ab \cos \gamma$ and c^2 on the right.
- In blue, the triangle *ABC*, on the left and on the right.
- In grey, auxiliary triangles, all congruent to *ABC*, an equal number (namely 2) both on the left and on the right.

The equality of areas on the left and on the right gives,

$$a^2 + b^2 = c^2 + 2ab\cos\gamma.$$

Obtuse Case

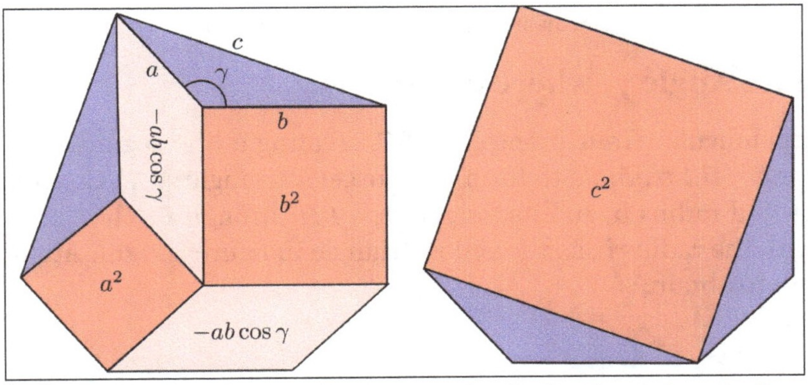

Proof of the law of cosines for obtuse angle γ by "cutting and pasting".

Figure cuts a hexagon in two different ways into smaller pieces, yielding a proof of the law of cosines in the case that the angle γ is obtuse. We have,

- In pink, the areas a^2, b^2, and $-2ab \cos \gamma$ on the left and c^2 on the right;
- In blue, the triangle *ABC* twice, on the left, as well as on the right.

The equality of areas on the left and on the right gives,

$$a^2 + b^2 - 2ab\cos(\gamma) = c^2.$$

The rigorous proof will have to include proofs that various shapes are congruent and therefore have equal area. This will use the theory of congruent triangles.

Using Geometry of the Circle

Using the geometry of the circle, it is possible to give a more geometric proof than using the Pythagorean theorem alone. Algebraic manipulations (in particular the binomial theorem) are avoided.

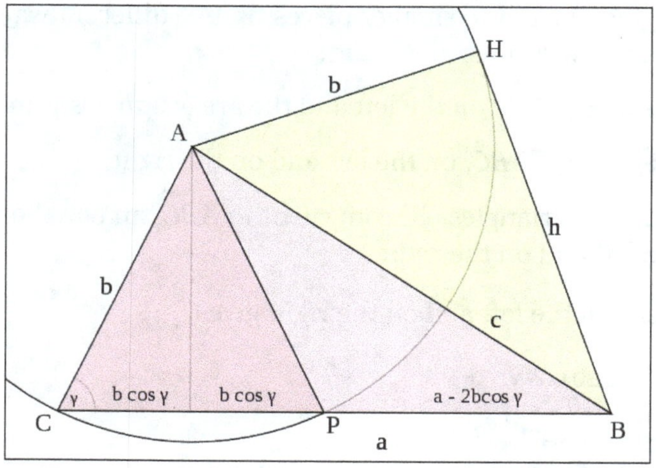

The triangle ABC (pink), an auxiliary circle (light blue) and an auxiliary right triangle (yellow).

Case of Acute Angle γ, where $a > 2b \cos \gamma$

Drop the perpendicular from A onto $a = BC$, creating a line segment of length $b \cos \gamma$. Duplicate the *right triangle* to form the isosceles triangle ACP. Construct the circle with center A and radius b, and its tangent $h = BH$ through B. The tangent h forms a right angle with the radius b, so the yellow triangle in figure is right. Apply the Pythagorean theorem to obtain,

$$c^2 = b^2 + h^2.$$

Then use the *tangent secant theorem*, which says that the square on the tangent through a point B outside the circle is equal to the product of the two lines segments (from B) created by any secant of the circle through B. In the present case: $BH^2 = BC \cdot BP$, or

$$h^2 = a(a - 2b\cos\gamma).$$

Substituting into the previous equation gives the law of cosines:

$$c^2 = b^2 + a(a - 2b\cos\gamma).$$

Note that h^2 is the power of the point B with respect to the circle. The use of the Pythagorean theorem and the tangent secant theorem can be replaced by a single application of the power of a point theorem.

Case of Acute Angle γ, where $a < 2b \cos \gamma$

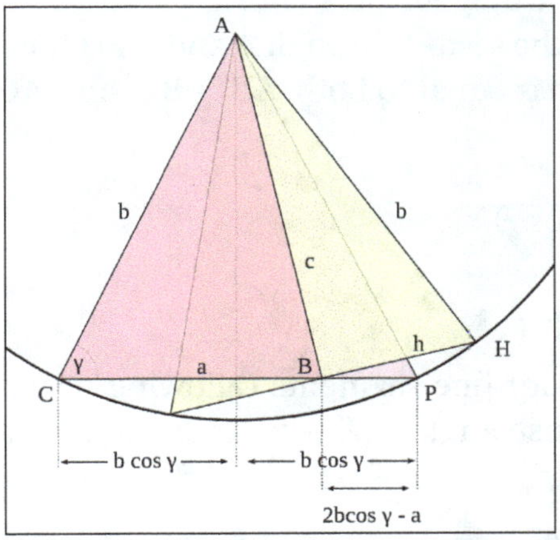

The triangle ABC (pink), an auxiliary circle (light blue) and two auxiliary right triangles (yellow).

Drop the perpendicular from A onto $a = BC$, creating a line segment of length $b \cos \gamma$. Duplicate the right triangle to form the isosceles triangle ACP. Construct the circle with center A and radius b, and a chord through B perpendicular to $c = AB$, half of which is $h = BH$. Apply the Pythagorean theorem to obtain

$$b^2 = c^2 + h^2.$$

Now use the *chord theorem*, which says that if two chords intersect, the product of the two line segments obtained on one chord is equal to the product of the two line segments obtained on the other chord. In the present case: $BH^2 = BC \cdot BP$, or

$$h^2 = a(2b \cos \gamma - a).$$

Substituting into the previous equation gives the law of cosines:

$$b^2 = c^2 + a(2b \cos \gamma - a).$$

Note that the power of the point B with respect to the circle has the negative value $-h^2$.

Case of Obtuse Angle γ

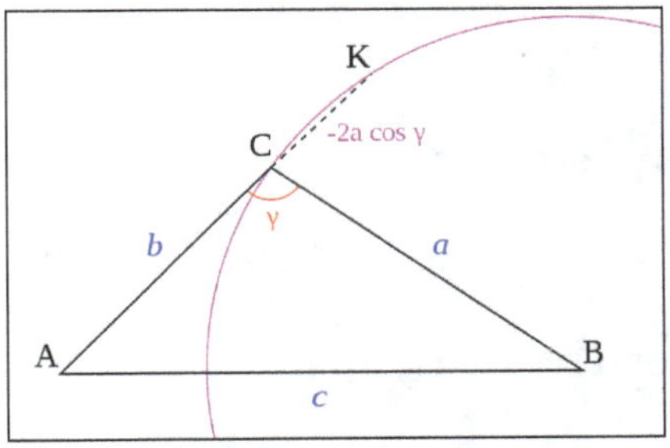

Proof of the law of cosines using the power of a point theorem.

This proof uses the power of a point theorem directly, without the auxiliary triangles obtained by constructing a tangent or a chord. Construct a circle with center *B* and radius *a*, which intersects the secant through *A* and *C* in *C* and *K*. The power of the point *A* with respect to the circle is equal to both $AB^2 - BC^2$ and *AC·AK*. Therefore,

$$\begin{aligned} c^2 - a^2 &= b(b + 2a\cos(\pi - \gamma)) \\ &= b(b - 2a\cos\gamma), \end{aligned}$$

which is the law of cosines.

Using algebraic measures for line segments (allowing negative numbers as lengths of segments) the case of obtuse angle ($CK > 0$) and acute angle ($CK < 0$) can be treated simultaneously.

Using the Law of Sines

By using the law of sines and knowing that the angles of a triangle must sum to 180 degrees, we have the following system of equations (the three unknowns are the angles):

$$\frac{c}{\sin\gamma} = \frac{b}{\sin\beta},$$

$$\frac{c}{\sin\gamma} = \frac{a}{\sin\alpha},$$

$$\alpha + \beta + \gamma = \pi.$$

Then, by using the third equation of the system, we obtain a system of two equations in two variables:

$$\frac{c}{\sin\gamma} = \frac{b}{\sin(\alpha + \gamma)},$$

$$\frac{c}{\sin\gamma} = \frac{a}{\sin\alpha},$$

where we have used the trigonometric property that the sine of a supplementary angle is equal to the sine of the angle.

Using the identity,

$$\sin(\alpha + \gamma) = \sin\alpha\cos\gamma + \sin\gamma\cos\alpha$$

leads to,

$$c(\sin\alpha\cos\gamma + \sin\gamma\cos\alpha) = b\sin\gamma,$$

$$c\sin\alpha = a\sin\gamma.$$

By dividing the whole system by cos γ, we have:

$$c(\sin\alpha + \tan\gamma\cos\alpha) = b\tan\gamma,$$

$$\frac{c\sin\alpha}{\cos\gamma} = a\tan\gamma,$$

$$\frac{c^2\sin^2\alpha}{\cos^2\gamma} = a^2\tan^2\gamma.$$

Hence, from the first equation of the system, we can obtain,

$$\frac{c\sin\alpha}{b - c\cos\alpha} = \tan\gamma$$

By substituting this expression into the second equation and by using,

$$1 + \tan^2\gamma = \frac{1}{\cos^2\gamma}$$

we can obtain one equation with one variable:

$$c^2\sin^2\alpha\left(1 + \frac{c^2\sin^2\alpha}{(b - c\cos\alpha)^2}\right) = a^2 \cdot \frac{c^2\sin^2\alpha}{(b - c\cos\alpha)^2}$$

By multiplying by $(b - c\cos\alpha)^2$, we can obtain the following equation:

$$(b - c\cos\alpha)^2 + c^2\sin^2\alpha = a^2.$$

This implies,

$$b^2 - 2bc\cos\alpha + c^2\cos^2\alpha + c^2\sin^2\alpha = a^2.$$

Recalling the Pythagorean identity, we obtain the law of cosines:

$$a^2 = b^2 + c^2 - 2bc\cos\alpha.$$

Isosceles Case

When $a = b$, i.e., when the triangle is isosceles with the two sides incident to the angle γ equal, the law of cosines simplifies significantly. Namely, because $a^2 + b^2 = 2a^2 = 2ab$, the law of cosines becomes,

$$\cos\gamma = 1 - \frac{c^2}{2a^2}$$

or,

$$c^2 = 2a^2(1 - \cos\gamma).$$

Analogue for Tetrahedra

An analogous statement begins by taking $\alpha, \beta, \gamma, \delta$ to be the areas of the four faces of a tetrahedron. Denote the dihedral angles by $\widehat{\beta\gamma}$ etc. Then,

$$\alpha^2 = \beta^2 + \gamma^2 + \delta^2 - 2\left(\beta\gamma\cos\left(\widehat{\beta\gamma}\right) + \gamma\delta\cos\left(\widehat{\gamma\delta}\right) + \delta\beta\cos\left(\widehat{\delta\beta}\right)\right).$$

Version Suited to Small Angles

When the angle, γ, is small and the adjacent sides, a and b, are of similar length, the right hand side of the standard form of the law of cosines can lose a lot of accuracy to numerical loss of significance. In situations where this is an important concern, a mathematically equivalent version of the law of cosines, similar to the haversine formula, can prove useful:

$$\begin{aligned} c^2 &= (a-b)^2 + 4ab\sin^2\left(\frac{\gamma}{2}\right) \\ &= (a-b)^2 + 4ab\,\text{haversin}(\gamma). \end{aligned}$$

In the limit of an infinitesimal angle, the law of cosines degenerates into the circular arc length formula, $c = a\,\gamma$.

In Spherical and Hyperbolic Geometry

Versions similar to the law of cosines for the Euclidean plane also hold on a unit sphere and in a hyperbolic plane. In spherical geometry, a triangle is defined by three points **u**, **v**, and **w** on the unit sphere, and the arcs of great circles connecting those points. If these great circles make angles A, B, and C with opposite sides a, b, c then the spherical law of cosines asserts that both of the following relationships hold:

$$\begin{aligned} \cos a &= \cos b\cos c + \sin b\sin c\cos A \\ \cos A &= -\cos B\cos C + \sin B\sin C\cos a. \end{aligned}$$

In hyperbolic geometry, a pair of equations are collectively known as the hyperbolic law of cosines. The first is,

$$\cosh a = \cosh b\cosh c - \sinh b\sinh c\cos A$$

where sinh and cosh are the hyperbolic sine and cosine, and the second is,

$$\cos A = -\cos B\cos C + \sin B\sin C\cosh a.$$

As in Euclidean geometry, one can use the law of cosines to determine the angles A, B, C from the knowledge of the sides a, b, c. In contrast to Euclidean geometry, the reverse is also possible in both non-Euclidean models: the angles A, B, C determine the sides a, b, c.

Unified Formula for Surfaces of Constant Curvature

Defining two functions $\cos_R$ and $\sin_R$ as,

$$\cos_R(x) = \cos(x/R) \quad \text{and} \quad \sin_R(x) = R \cdot \sin(x/R)$$

allows to unify the formulae for plane, sphere and pseudosphere into:

$$\cos_R(BC) = \cos_R(AB) \cdot \cos_R(AC) + \frac{1}{R^2} \sin_R(AB) \cdot \sin_R(AC) \cdot \cos(\widehat{BAC}).$$

In this notation R is a complex number, representing the surface's radius of curvature.

- For $R \in \mathbb{R}$ the surface is a sphere of radius R, and its constant curvature equals $1/R^2$.
- For $R = iR' : R' \in \mathbb{R}$ the surface is a pseudosphere of (imaginary) radius R, with constant curvature equal to $1/R^2 = -1/R'^2$.
- For $R \to \infty$: The surface tends to a Euclidean plane, with constant zero curvature.

Verifying the Formula for non-Euclidean Geometry

In the first two cases, $\cos_R$ and $\sin_R$ are well-defined over the whole complex plane for all $R \neq 0$, and retrieving former results is straightforward.

Hence, for a sphere of radius 1

$$\cos(BC) = \cos(AB) \cdot \cos(AC) + \sin(AB) \cdot \sin(AC) \cdot \cos(\widehat{BAC}).$$

Likewise, for a pseudosphere of radius i

$$\cosh(BC) = \cosh(AB) \cdot \cosh(AC) - \sinh(AB) \cdot \sinh(AC) \cdot \cos(\widehat{BAC}).$$

Indeed, $\cosh(x) = \cos(x/i)$ and $\sinh(x) = i \cdot \sin(x/i)$.

Verifying the Formula in the Limit of Euclidean Geometry

In the Euclidean plane the appropriate limits for the above equation must be calculated:

$$\cos_R(x) = \cos(x/R) = 1 - \frac{1}{2} \cdot \frac{x^2}{R^2} + o\left(\frac{1}{R^4}\right)$$

and

$$\sin_R(x) = R \cdot \sin(x/R) = x + o\left(\frac{1}{R^3}\right)$$

Applying this to the general formula for a finite R yields:

$$1-\frac{BC^2}{2R^2}+o\left(\frac{1}{R^4}\right)=\left(1-\frac{AB^2}{2R^2}+o\left(\frac{1}{R^4}\right)\right)\cdot\left(1-\frac{AC^2}{2R^2}+o\left(\frac{1}{R^4}\right)\right)$$
$$+\frac{1}{R^2}\left(AB+o\left(\frac{1}{R^3}\right)\right)\cdot\left(AC+o\left(\frac{1}{R^3}\right)\right)\cdot\cos(\widehat{BAC})$$

Collecting terms, multiplying with $-2R^2$, and taking $R\to\infty$ yields the expected formula:

$$BC^2=AB^2+AC^2-2\cdot AB\cdot AC\cdot\cos(\widehat{BAC}).$$

Spherical Law of Cosines

In spherical trigonometry, the law of cosines (also called the cosine rule for sides) is a theorem relating the sides and angles of spherical triangles, analogous to the ordinary law of cosines from plane trigonometry.

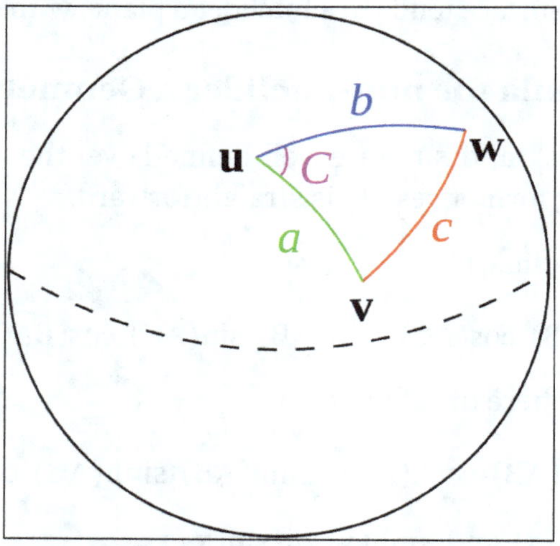

Spherical triangle solved by the law of cosines.

Given a unit sphere, a "spherical triangle" on the surface of the sphere is defined by the great circles connecting three points u, v, and w on the sphere (shown at right). If the lengths of these three sides are *a* (from u to v), *b* (from u to w), and c (from v to w), and the angle of the corner opposite *c* is *C*, then the (first) spherical law of cosines states:

$$\cos c=\cos a\cos b+\sin a\sin b\cos C$$

Since this is a unit sphere, the lengths *a*, *b*, and *c* are simply equal to the angles (in radians) subtended by those sides from the center of the sphere. (For a non-unit sphere, the lengths are the subtended angles times the radius, and the formula still holds if a, b and c are reinterpreted as the subtended angles). As a special case, for $C = \frac{\pi}{2}$ then $\cos C = 0$, and one obtains the spherical analogue of the Pythagorean theorem:

$$\cos c = \cos a \cos b$$

If the law of cosines is used to solve for c, the necessity of inverting the cosine magnifies rounding errors when c is small. In this case, the alternative formulation of the law of haversines is preferable.

A variation on the law of cosines, the second spherical law of cosines, (also called the cosine rule for angles) states:

$$\cos C = -\cos A \cos B + \sin A \sin B \cos c$$

where A and B are the angles of the corners opposite to sides a and b, respectively. It can be obtained from consideration of a spherical triangle dual to the given one.

Proofs:

First proof:

Let u, v, and w denote the unit vectors from the center of the sphere to those corners of the triangle. The angles and distances do not change if the coordinate system is rotated, so we can rotate the coordinate system so that u is at the north pole and v is somewhere on the prime meridian (longitude of 0). With this rotation, the spherical coordinates for v are $(r, \theta, \phi) = (1, a, 0)$, where θ is the angle measured from the north pole not from the equator, and the spherical coordinates for w are, $(r, \theta, \phi) = (1, b, C)$. The Cartesian coordinates for v are $(x, y, z) = (\sin a, 0, \cos a)$ and the Cartesian coordinates for w are $(x, y, z) = (\sin b \cos C, \sin b \sin C, \cos b)$. The value of $\cos c$ is the dot product of the two Cartesian vectors, which is $\sin a \sin b \cos C + \cos a \cos b$.

Second proof:

Let $\mathbf{u}$, $\mathbf{v}$, and $\mathbf{w}$ denote the unit vectors from the center of the sphere to those corners of the triangle. We have $\mathbf{u} \cdot \mathbf{u} = 1$, $\mathbf{v} \cdot \mathbf{w} = \cos c$, $\mathbf{u} \cdot \mathbf{v} = \cos a$, and $\mathbf{u} \cdot \mathbf{w} = \cos b$. The vectors $\mathbf{u} \times \mathbf{v}$ and $\mathbf{u} \times \mathbf{w}$ have lengths $\sin a$ and $\sin b$ respectively and the angle between them is C, so,

$$\sin a \sin b \cos C = (\mathbf{u} \times \mathbf{v}) \cdot (\mathbf{u} \times \mathbf{w}) = (\mathbf{u} \cdot \mathbf{u})(\mathbf{v} \cdot \mathbf{w}) - (\mathbf{u} \cdot \mathbf{v})(\mathbf{u} \cdot \mathbf{w})$$

$$= \cos c - \cos a \cos b,$$

using cross products, dot products, and the Binet–Cauchy identity

$$(\mathbf{p} \times \mathbf{q}) \cdot (\mathbf{r} \times \mathbf{s}) = (\mathbf{p} \cdot \mathbf{r})(\mathbf{q} \cdot \mathbf{s}) - (\mathbf{p} \cdot \mathbf{s})(\mathbf{q} \cdot \mathbf{r}).$$

Rearrangements

The first and second spherical laws of cosines can be rearranged to put the sides (*a*, *b*, *c*) and angles (*A*, *B*, *C*) on opposite sides of the equations:

$$\cos C = \frac{\cos c - \cos a \cos b}{\sin a \sin b}$$

$$\cos c = \frac{\cos C + \cos A \cos B}{\sin A \sin B}$$

Planar Limit: Small Angles

For *small* spherical triangles, i.e. for small *a*, *b*, and *c*, the spherical law of cosines is approximately the same as the ordinary planar law of cosines,

$$c^2 \approx a^2 + b^2 - 2ab\cos C.$$

To prove this, we will use the small-angle approximation obtained from the Maclaurin series for the cosine and sine functions:

$$\cos a = 1 - \frac{a^2}{2} + O\left(a^4\right), \sin a = a + O\left(a^3\right)$$

Substituting these expressions into the spherical law of cosines nets:

$$1 - \frac{c^2}{2} + O\left(c^4\right) = 1 - \frac{a^2}{2} - \frac{b^2}{2} + \frac{a^2b^2}{4} + O\left(a^4\right) + O\left(b^4\right) + \cos(C)\left(ab + O\left(a^3b\right)\right.$$
$$\left. + O\left(ab^3\right) + O\left(a^3b^3\right)\right)$$

or after simplifying:

$$c^2 = a^2 + b^2 - 2ab\cos C + O\left(c^4\right) + O\left(a^4\right) + O\left(b^4\right) + O\left(a^2b^2\right) + O\left(a^3b\right)\cdot$$
$$+ O\left(ab^3\right) + O\left(a^3b^3\right).$$

The big O terms for *a* and *b* are dominated by $O(a^4) + O(b^4)$ as *a* and *b* get small, so we can write this last expression as:

$$c^2 = a^2 + b^2 - 2ab\cos C + O\left(a^4\right) + O\left(b^4\right) + O\left(c^4\right).$$

Hyperbolic Law of Cosines

In hyperbolic geometry, the "law of cosines" is a pair of theorems relating the sides and angles of triangles on a hyperbolic plane, analogous to the planar law of cosines from plane trigonometry, or the spherical law of cosines in spherical trigonometry. It can also be related to the relativisic velocity addition formula.

Describing relations of hyperbolic geometry, it was shown by Franz Taurinus that the spherical law of cosines can be related to spheres of imaginary radius, thus he arrived at the hyperbolic law of cosines in the form:

$$A = \arccos\frac{\cos\left(\alpha\sqrt{-1}\right)-\cos\left(\beta\sqrt{-1}\right)\cos\left(\gamma\sqrt{-1}\right)}{\sin\left(\beta\sqrt{-1}\right)\sin\left(\gamma\sqrt{-1}\right)}$$

which was also shown by Nikolai Lobachevsky:

$$\cos A\sin b\sin c-\cos b\cos c=\cos a;\quad [a,b,c]\to\left[a\sqrt{-1},b\sqrt{-1},c\sqrt{-1}\right]$$

Ferdinand Minding gave it in relation to surfaces of constant negative curvature:

$$\cos a\sqrt{k}=\cos b\sqrt{k}\cdot\cos c\sqrt{k}+\sin b\sqrt{k}\cdot\sin c\sqrt{k}\cdot\cos A$$

as did Delfino Codazzi:

$$\cos\beta p\left(\frac{a}{r}\right)p\left(\frac{s}{r}\right)=q\left(\frac{a}{r}\right)q\left(\frac{s}{r}\right)-q\left(\frac{\lambda}{r}\right),\quad\left[\frac{e^{t}-e^{-t}}{2}=p(t),\frac{e^{t}+e^{-t}}{2}=q(t)\right]$$

The relation to relativity using rapidity was shown by Arnold Sommerfeld and Vladimir Varićak.

Take a hyperbolic plane whose Gaussian curvature is $-\frac{1}{k^2}$. Then given a hyperbolic triangle ABC with angles α,β,γ and side lengths $BC=a$, $AC=b$, and $AB=c$, the following two rules hold:

$$\cosh\frac{a}{k}=\cosh\frac{b}{k}\cosh\frac{c}{k}-\sinh\frac{b}{k}\sinh\frac{c}{k}\cos\alpha,$$

considering the sides, while,

$$\cos\alpha=-\cos\beta\cos\gamma+\sin\beta\sin\gamma\cosh\frac{a}{k},$$

Christian Houzel indicates that the hyperbolic law of cosines implies the angle of parallelism in the case of an ideal hyperbolic triangle:

> When $\alpha=0$, that is when the vertex "A" is rejected to infinity and the sides "BA"

and “CA” are ”parallel”, the first member equals 1; let us suppose in addition that $\gamma = \pi/2$ so that $\cos\gamma = 0$ and $\sin\gamma = 1$. The angle at ”B” takes a value β given by $1 = \sin\beta\cosh(a/k)$; this angle was later called ”angle of parallelism” and Lobachevsky noted it by “F(a)” or Π(“a”).

Hyperbolic Law of Haversines

In cases where “a/k” is small, and being solved for, the numerical precision of the standard form of the hyperbolic law of cosines will drop due to rounding errors, for exactly the same reason it does in the Spherical law of cosines. The hyperbolic version of the law of haversines can prove useful in this case:

$$\sinh^2\frac{a}{2k} = \sinh^2\frac{b-c}{2k} + \sinh\frac{b}{k}\sinh\frac{c}{k}\sin^2\frac{\alpha}{2},$$

Relativistic Velocity Addition via Hyperbolic Law of Cosines

Setting $\left[\frac{a}{k},\frac{b}{k},\frac{c}{k}\right] = [\xi,\eta,\zeta]$ in $\cosh\frac{a}{k} = \cosh\frac{b}{k}\cosh\frac{c}{k} - \sinh\frac{b}{k}\sinh\frac{c}{k}\cos\alpha$, and by using hyperbolic identities in terms of the hyperbolic tangent, the hyperbolic law of cosines can be written:

$$\cosh\xi = \cosh\eta\cosh\zeta - \sinh\eta\sinh\zeta\cos\alpha$$

$$\Rightarrow \frac{1}{\sqrt{1-\tanh^2\xi}} = \frac{1}{\sqrt{1-\tanh^2\eta}}\frac{1}{\sqrt{1-\tanh^2\zeta}} - \frac{\tanh\eta}{\sqrt{1-\tanh^2\eta}}\frac{\tanh\zeta}{\sqrt{1-\tanh^2\zeta}}\cos\alpha$$

$$\Rightarrow \tanh\xi = \frac{\sqrt{-\tanh^2\zeta - \tanh^2\eta + 2\tanh\eta\tanh\zeta\cos\alpha + (\tanh\eta\tanh\zeta\sin\alpha)^2}}{1-\tanh\eta\tanh\zeta\cos\alpha}$$

In comparison, the velocity addition formulas of special relativity for the x and y-directions as well as under an arbitrary angle α, where v is the relative velocity between two inertial frames, u the velocity of another object or frame, and c the speed of light, is given by,

$$[U_x, U_y] = \left[\frac{u_x - v}{1-\frac{v}{c^2}u_x}, \frac{u_y\sqrt{1-\frac{v^2}{c^2}}}{1-\frac{v}{c^2}u_x}\right]$$

$$U^2 = U_x^2 + U_y^2,\ u^2 = u_x^2 + u_y^2,\ \tan\alpha = \frac{u_y}{u_x}$$

$$\Rightarrow \qquad U = \frac{\sqrt{-u^2 - v^2 + 2vu\cos\alpha + \left(\frac{vu\sin\alpha}{c}\right)^2}}{1 - \frac{v}{c^2}u\cos\alpha}$$

It turns out that this result corresponds to the hyperbolic law of cosines - by identifying $[\xi, \eta, \zeta]$ with relativistic rapidities

$$\left(\left[\frac{U}{c}, \frac{v}{c}, \frac{u}{c}\right] = [\tanh\xi, \tanh\eta, \tanh\zeta]\right),$$

the equations in hyperbolic law of cosines assume the form:

$$\cosh\xi = \cosh\eta\cosh\zeta - \sinh\eta\sinh\zeta\cos\alpha$$

$$\Rightarrow \frac{1}{\sqrt{1-\frac{U^2}{c^2}}} = \frac{1}{\sqrt{1-\frac{v^2}{c^2}}}\frac{1}{\sqrt{1-\frac{u^2}{c^2}}} - \frac{v/c}{\sqrt{1-\frac{v^2}{c^2}}}\frac{u/c}{\sqrt{1-\frac{u^2}{c^2}}}\cos\alpha$$

$$\Rightarrow \qquad U = \frac{\sqrt{-u^2 - v^2 + 2vu\cos\alpha + \left(\frac{vu\sin\alpha}{c}\right)^2}}{1 - \frac{v}{c^2}u\cos\alpha}.$$

LAW OF TANGENTS

The law of tangent (Law of Tan) describes the relation between difference and sum of sides of a right triangle and tangents of half of the difference and sum of corresponding angles. It represents the relationship between the tangent of two angles of a triangle and the length of the opposite sides. The law of tangents is also applied to a non-right triangle and it is equally as powerful like the law of sines and the law of cosines. It can be used to find the remaining parts of a triangle if two angles and one side or two sides and one angle are given which are referred to as side-angle-side(SAS) and angle-side-angle(ASA) , from congruence of triangles concept.

To understand the law of tangents in a better way, we will need three pieces of information for a general triangle even it may or may not be a triangle. The four cases involved are:

- Two sides and one opposite angle.
- One side and two angles.

- Three Sides.
- Two sides and the angle between them.

Formulas for Laws of Tangents

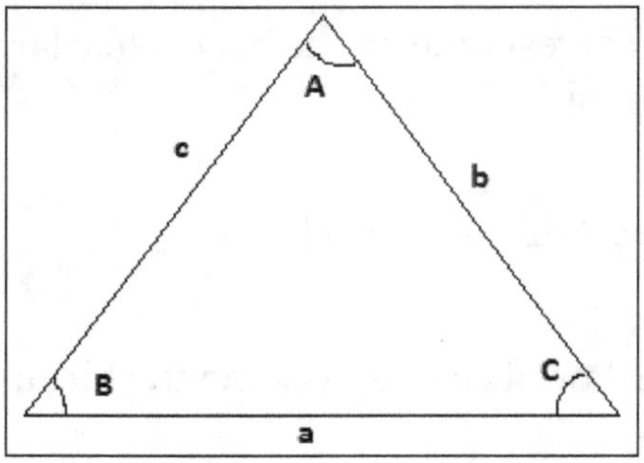

Let us assume a right triangle ABC in which side opposite to ∠A, ∠B, and ∠C are a, b and c respectively. Then, according to the laws of tangent, we have the following three relations:

$$\frac{a-b}{a+b}=\frac{\tan\left(\frac{A-B}{2}\right)}{\tan\left(\frac{A+B}{2}\right)}$$

Similarly for other sides,

$$\frac{b-c}{b+c}=\frac{\tan\left(\frac{B-C}{2}\right)}{\tan\left(\frac{B+C}{2}\right)}$$

$$\frac{c-a}{c+a}=\frac{\tan\left(\frac{C-A}{2}\right)}{\tan\left(\frac{C+A}{2}\right)}$$

Since tan (-θ) = -tan θ for any angle θ, we can switch the order of letters in the above law of tangents formulas and it is rewritten as:

$$\frac{b-a}{b+a}=\frac{\tan\left(\frac{B-A}{2}\right)}{\tan\left(\frac{B+A}{2}\right)}$$

Similarly for other sides,

$$\frac{c-b}{c+b}=\frac{\tan\left(\frac{C-B}{2}\right)}{\tan\left(\frac{C+B}{2}\right)}$$

$$\frac{a-c}{a+c}=\frac{\tan\left(\frac{A-C}{2}\right)}{\tan\left(\frac{A+C}{2}\right)}$$

The formulas

$$\frac{a-b}{a+b}=\frac{\tan\left(\frac{A-B}{2}\right)}{\tan\left(\frac{A+B}{2}\right)},\ \frac{b-c}{b+c}=\frac{\tan\left(\frac{B-C}{2}\right)}{\tan\left(\frac{B+C}{2}\right)},$$

$$\text{and } \frac{c-a}{c+a}=\frac{\tan\left(\frac{C-A}{2}\right)}{\tan\left(\frac{C+A}{2}\right)},$$

are used when a>b, b>c, and c>a, and the formulas

$$\frac{b-a}{b+a}=\frac{\tan\left(\frac{B-A}{2}\right)}{\tan\left(\frac{B+A}{2}\right)},\ \frac{c-b}{c+b}=\frac{\tan\left(\frac{C-B}{2}\right)}{\tan\left(\frac{C+B}{2}\right)}$$

$$\text{and } \frac{a-c}{a+c}=\frac{\tan\left(\frac{A-C}{2}\right)}{\tan\left(\frac{A+C}{2}\right)}$$

are used when b>a, c>b and a>c.

Laws of Tangent Proof

To Prove:

$$\frac{a-b}{a+b}=\frac{\tan\left(\frac{A-B}{2}\right)}{\tan\left(\frac{A+B}{2}\right)}$$

Proof: From the law of Sine,

$$\frac{a}{\sin A} = \frac{b}{\sin B} = \frac{c}{\sin C}$$

Use first and second relation,

$$\frac{a}{\sin A} = \frac{b}{\sin B} = k,$$

$$a = k \sin A \text{ and } b = k \sin B$$

From this,

$$a - b = k(\sin A - \sin B)$$

$$a + b = k(\sin A + \sin B)$$

So, we get,

$$\frac{a-b}{a+b} = \frac{\sin A - \sin B}{\sin A + \sin B}$$

Identity Formulas for Sine are:

$$\sin A - \sin B = 2\cos\frac{A+B}{2}\sin\frac{A-B}{2}$$

$$\sin A + \sin B = 2\sin\frac{A+B}{2}\cos\frac{A-B}{2}$$

Substitute those formulas in equation

$$\frac{a-b}{a+b} = \frac{\tan\left(\frac{A-B}{2}\right)}{\tan\left(\frac{A+B}{2}\right)}$$

we get,

$$\frac{a-b}{a+b} = \frac{2\cos\frac{A+B}{2}\sin\frac{A-B}{2}}{2\sin\frac{A+B}{2}\cos\frac{A-B}{2}}$$

$$= \frac{\tan\frac{A-B}{2}}{\tan\frac{A+B}{2}}$$

Hence Proved.

Practice Problem

Example:

Solve the triangle $\triangle ABC$ given a=5, b=3 and $\angle C=96°$ and find the value of A – B.

Solution:

We know that,

$$\angle A + \angle B + \angle C = 180°$$
$$\angle A + \angle B = 180° - \angle C = 180° - 96° = 84°$$

By law of tangents,

for a triangle ABC with sides a, b and c respective to the angles A , B and C is given by,

$$\frac{a-b}{a+b} = \frac{\tan\left(\frac{A-B}{2}\right)}{\tan\left(\frac{A+B}{2}\right)}$$

Therefore,

$$\Rightarrow \frac{5-3}{5+3} = \frac{\tan\frac{1}{2}(A-B)}{\tan\frac{1}{2}(84°)} = \tan\frac{1}{2}(A-B) = \frac{2}{8}\tan 42° = 0.2251$$

$$\Rightarrow \frac{1}{2}(A-B) = 12.7°$$
$$A - B = 25.4°$$

References

- Law-of-sines: varsitytutors.com, Retrieved 13 April, 2020
- Pickover, Clifford A. (2009). The Math Book: From Pythagoras to the 57th Dimension, 250 Milestones in the History of Mathematics. Sterling Publishing Company, Inc. p. 106. ISBN 9781402757969
- Anderson, James W. (2005). Hyperbolic geometry (2nd ed.). London: Springer. ISBN 1-85233-934-9
- Law-of-tangents, maths: byjus.com, Retrieved 05 May, 2020

Permissions

All chapters in this book are published with permission under the Creative Commons Attribution Share Alike License or equivalent. Every chapter published in this book has been scrutinized by our experts. Their significance has been extensively debated. The topics covered herein carry significant information for a comprehensive understanding. They may even be implemented as practical applications or may be referred to as a beginning point for further studies.

We would like to thank the editorial team for lending their expertise to make the book truly unique. They have played a crucial role in the development of this book. Without their invaluable contributions this book wouldn't have been possible. They have made vital efforts to compile up to date information on the varied aspects of this subject to make this book a valuable addition to the collection of many professionals and students.

This book was conceptualized with the vision of imparting up-to-date and integrated information in this field. To ensure the same, a matchless editorial board was set up. Every individual on the board went through rigorous rounds of assessment to prove their worth. After which they invested a large part of their time researching and compiling the most relevant data for our readers.

The editorial board has been involved in producing this book since its inception. They have spent rigorous hours researching and exploring the diverse topics which have resulted in the successful publishing of this book. They have passed on their knowledge of decades through this book. To expedite this challenging task, the publisher supported the team at every step. A small team of assistant editors was also appointed to further simplify the editing procedure and attain best results for the readers.

Apart from the editorial board, the designing team has also invested a significant amount of their time in understanding the subject and creating the most relevant covers. They scrutinized every image to scout for the most suitable representation of the subject and create an appropriate cover for the book.

The publishing team has been an ardent support to the editorial, designing and production team. Their endless efforts to recruit the best for this project, has resulted in the accomplishment of this book. They are a veteran in the field of academics and their pool of knowledge is as vast as their experience in printing. Their expertise and guidance has proved useful at every step. Their uncompromising quality standards have made this book an exceptional effort. Their encouragement from time to time has been an inspiration for everyone.

The publisher and the editorial board hope that this book will prove to be a valuable piece of knowledge for students, practitioners and scholars across the globe.

Index

A
Acute Angle, 35-36, 212, 214, 216-218
Angle of Depression, 30
Angle Sum Identities, 177, 180
Arc Length, 43-44, 185, 220

B
Binomial Coefficients, 61, 199
Binomial Theorem, 173, 199-200, 213, 216

C
Cartesian Coordinate System, 36, 40-41, 210
Cartesian Plane, 37, 50
Catalan Constant, 76, 87, 96
Chain Rule, 117, 173
Constant Coefficients, 56
Cosecant Function, 28, 70-73, 75-77, 79-80, 87-88, 133, 138
Cosine Function, 2, 29, 50-52, 54-55, 57, 59, 62-63, 80, 82, 90, 98, 121, 130, 138
Cotangent Function, 28, 90-93, 95-97, 99-100, 132, 139
Cyclic Quadrilateral, 164, 213

D
Differential Equations, 33, 35, 56, 173
Distance Formula, 210

E
Empty Product, 167
Euler Numbers, 85

F
Fermat Primes, 21, 25, 55, 74, 84, 93, 162
Finite Summation, 58, 77, 87, 96
Fourier Series, 43, 48

G
Gamma Function, 43, 49

H
Holomorphic Function, 49
Hyperbolic Functions, 62, 80, 90, 100
Hypergeometric Function, 53, 96

I
Imaginary Unit, 47
Indefinite Integrals, 57, 76, 86, 95
Indefinite Integration, 57, 76, 86, 95
Infinite Series, 3, 35, 48, 127
Inverse Cosine, 113-115, 130
Inverse Function, 8, 37, 61, 79, 89, 99, 109, 113, 115-117, 119, 126, 129
Inverse Substitution, 8-9

J
Jacobi Elliptic Functions, 54, 72, 82, 92

K
Kronecker Delta, 75, 85, 94

L
Laurent Series, 75, 95, 127
Law of Cosines, 2, 5, 203, 208-209, 212-220, 222-227
Law of Sines, 2, 5, 203-204, 206-208, 218, 227
Law of Tangents, 6, 227-228, 231
Legendre Polynomial, 120
Limit Representation, 86
Line Segment, 35, 71, 81, 91, 126, 211-212, 216-217

M
Multiplicative Inverse, 109, 113, 119, 126

N
Natural Logarithm, 64, 126

O
Ordinary Differential Equation, 56, 75, 85, 94, 174

P
Partial Fraction, 48
Periodic Function, 37, 56, 74, 84, 94

Pochhammer Symbol, 111, 120
Polynomial Function, 16, 18, 96
Power Series, 118-119, 172-174
Principal Value, 37, 103, 109, 113, 119, 126, 152
Product Identities, 4, 183
Product Representation, 57, 76, 86
Pythagorean Identities, 3, 135-136, 176, 180-181
Pythagorean Theorem, 3, 5, 10, 62, 129, 158, 169-174, 177, 208-209, 212-214, 216-217, 223

R
Ratio Identities, 135, 175-176
Rational Function, 78, 88, 116, 118
Rational Numbers, 55
Real Line, 53, 67, 126
Real Numbers, 20, 42, 81, 115, 152, 158
Reciprocal Identities, 135
Reference Angle, 52, 66, 103
Reference Triangle, 129
Right Angle Triangle, 6, 8

S
Secant Function, 28, 80, 82-83, 85-87, 89-90, 121-122, 124, 132, 139
Series Representation, 56, 75, 85, 95
Sine Identities, 137, 183
Sine Rule, 5
Spherical Trigonometry, 1, 222, 225
Square Roots, 55, 74, 84, 93, 140, 166, 181
Substitution Rule, 8-9

T
Tangent Function, 2, 28, 63-64, 69, 91-92, 115, 131, 138
Tangent Line, 64, 116, 124
Taylor Series, 42, 118, 120
Trigonometric Functions, 1-3, 20-23, 25-26, 29-30, 33-36, 38-39, 58-59, 62, 71, 79-81, 98, 100, 105, 114, 123, 126, 129, 133, 140, 172-174, 185
Trigonometric Ratios, 6-7, 30-31, 102, 156
Trigonometric Substitution, 8, 12-14
Triple Angle Formulas, 199, 201

U
Unit Circle, 1, 3, 17-18, 35-37, 41, 44, 51, 64, 71, 81, 91-92, 129, 140, 170-173

W
Wolfram Language, 21, 23, 114, 119, 126

www.ingramcontent.com/pod-product-compliance
Lightning Source LLC
LaVergne TN
LVHW080618210926
848485LV00003B/177